PARTICLES OF TRUTH

PARTICLES OF TRUTH

A STORY OF DISCOVERY, CONTROVERSY,
AND THE FIGHT FOR HEALTHY AIR

C. ARDEN POPE III AND DOUGLAS W. DOCKERY

FOREWORD BY GINA MCCARTHY

THE MIT PRESS
CAMBRIDGE, MASSACHUSETTS
LONDON, ENGLAND

The MIT Press would like to thank the anonymous peer reviewers who provided comments on drafts of this book. The generous work of academic experts is essential for establishing the authority and quality of our publications. We acknowledge with gratitude the contributions of these otherwise uncredited readers.

This book was set in Stone Serif and Stone Sans by Jen Jackowitz. Printed and bound in the United States of America.

Library of Congress Cataloging-in-Publication Data is available.

ISBN: 978-0-262-55167-0

10 9 8 7 6 5 4 3 2 1

Dedicated to our grandchildren and great-grandchildren—that they would live in a world with healthier air.

CONTENTS

FOREWORD

Gina McCarthy

As someone who spent over forty years protecting public health and the environment, I can unequivocally say that Arden Pope and Doug Dockery are unsung superheroes in the field of air pollution research. So, when they sent me their book—*Particles of Truth: A Story of Discovery, Controversy, and the Fight for Healthy Air*—I immediately picked it up and couldn't put it down—and neither will you.

Particles of Truth is *not* a new study on the well-known human health effects of air pollution. At its heart, it is a real-life account of the groundbreaking science accumulated over the past four decades. It tells the story of the hard work that went into pulling together the body of evidence that has underpinned new policies, regulations, practices, and technologies that have successfully reduced air pollution and saved countless lives.

It is also a testament to the ingenuity, creativity, courage, and persistence of Doug, Arden, and the many dedicated research scientists across the world who have strived for decades to understand and document the connection between air pollution and health. As you will read, these individuals are consummate professionals who have spent their lives gathering and analyzing data, testing and retesting new methodologies, and expanding the body of research needed to connect the dots between air pollution and health with the precision that science demands.

Figure 1
Gina McCarthy. *Source:* Rose Lincoln, staff photographer, Harvard University.

This book brings back memories of pivotal moments in history when politicians were waging war against researchers with accusations of "secret science" and calls for the full disclosure of all personal health records used in groundbreaking efforts like the Six Cities study. Many of the researchers were called upon to stand up and defend their work under extremely challenging circumstances. Time and again they mustered the courage to effectively communicate the results of their research in the face of unwarranted and vitriolic criticism from naysayers and special interest groups that have always raised doubts and seeded confusion to preserve the profits of fossil fuel companies.

But *Particles of Truth* doesn't just give readers a look at the research and methods; it gives us the simplest and most compelling reminder of why this work matters to all of us. Air pollution destroys lives and livelihoods. The authors share a close—and painful—glimpse into the millions of innocent victims who die prematurely every year by introducing us to Ella, a child who lived in a flat in London located just feet from a congested major highway. Ella was seven years old when she had her first asthma attack, and only nine years old when she could not survive her twenty-seventh attack. We meet Rosamund, her mother, who was heartbroken over Ella's death, and how learning of the likelihood that air pollution was a large contributor to Ella's illness motivated her to work tirelessly for clean air in her daughter's memory.

Readers will leave with a solid understanding of why we must stop putting the interests of those profiting from fossil fuels over the health and well-being of our children and why clean energy is the only real path to protecting our health and climate. But, ultimately, this book is a celebration of science and a celebration of the superhero researchers who saved countless lives by working for decades to provide the evidence needed to fight for healthy air.

Gina McCarthy
1st White House National Climate Advisor and 13th Administrator of the U.S. Environmental Protection Agency

PROLOGUE

On the morning of November 2, 1989, Doug Dockery and Arden Pope met in person for the first time at the Harvard School of Public Health in Boston. Arden had just arrived on a red-eye flight from Utah. He was tired but excited to be there. Officially, he came to present a seminar to the Department of Environmental Science and Physiology about air pollution and respiratory health. Most importantly, he came to seek help and guidance from Doug.

Although both Arden and Doug were investigating the health effects of air pollution, they had arrived at this point in their careers from very different roots, and with different ideological perspectives about environmental policy.

Arden grew up in a family of ranchers and farmers in Wyoming and Idaho. His neighbors were good, hard-working people who loved their rural communities and a sense of independent living. The air was clean, and they were far removed from major polluting industries. Most of his family, friends, and neighbors were ideologically and politically conservative with libertarian perspectives. Arden remembers the passing of the Clean Air Act and the creation of the Environmental Protection Agency (EPA) in 1970 while in high school. These actions were not locally popular. The Clean Air Act and the formation of the EPA were considered political evils.

After graduating from high school in Idaho, Arden attended Brigham Young University. Motivated by his farm-boy background and a budding interest in applied econometrics and statistics, he earned a bachelor of science degree in agricultural economics. Yet he felt strangely unsatisfied. He had not been a highly dedicated or outstanding student, but now he yearned for more education. He enrolled in a graduate program at Iowa State University, where he became a dedicated student and researcher, receiving his PhD in 1981. His training and research interests included statistical, econometric, and mathematical programming analyses applied to natural resources and environmental issues—mostly related to agriculture. After two years as an assistant professor of agricultural economics at Texas A&M University, he returned to Brigham Young University in 1984, in large part to be closer to family.

Shortly after returning to Utah, Arden recognized a unique opportunity to conduct research dealing with the health effects of air pollution. This research took advantage of a "natural experiment," potentially revealing events that occur without the planning and control of scientists. The local steel mill, Geneva Steel, in Utah Valley had been closed for about a year and then reopened. Arden conducted statistical analysis of air pollution and health data for periods before, during, and after the closure. The results implicated air pollution from the steel mill as substantially contributing to children's hospital admissions for respiratory disease. These results were published in the *American Journal of Public Health*.[1]

When the research became public, the owners, operators, and political supporters of the steel mill unleashed a barrage of criticism toward Arden. They characterized his research as not using standard epidemiologic methods. They argued that since he was an agricultural economist, not an epidemiologist, he was not qualified or capable of conducting such research. Criticism of Arden's research was further stoked in early September when Geneva Steel reported in a news conference that a consultant medical scientist they had hired disputed Arden's research findings. The consultant claimed the increased respiratory hospital admissions of children were due to a virus and not air pollution.[2]

Arden was in a dilemma. He was a young, somewhat naive professor who had recognized a novel local, natural experiment as an opportunity to conduct a unique air pollution study, a side project to his main studies.

Now, he was under assault by a major local industry and its supporters and by a hired gun consultant. Arden needed either to bow out of this line of research or to more fully commit his efforts to understanding the health effects of air pollution. He was convinced that researching air pollution and health was important, and was willing to shift his academic priorities. But he needed additional help, guidance, and training. He needed quality collaborators with expertise in medicine, public health, epidemiology, atmospheric science, and related fields. How could he find what he needed?

Doug Dockery was born in Washington, DC, just a few blocks from the U.S. Capitol, and grew up just outside of Washington. Many of his neighbors worked in the military and government agencies. They were keenly aware of and appreciated the impact of the federal government on people's lives. During the 1960s, when Doug attended high school and college, much of the country was mobilized by President Kennedy's challenge to land a man on the moon by the end of the decade. There was a strong core belief in science and engineering and optimistic expectations for new government agencies such as NASA.

Doug attended high school across the street from NASA's Goddard Space Flight Center. He worked summers in high school as a laboratory assistant in programs funded by the National Science Foundation (NSF) at the nearby University of Maryland. In college, he worked part-time in laboratories designing and building rocket-borne and satellite sensors of the atmosphere. Upon graduation from the University of Maryland in 1969, he was accepted into the meteorology graduate program at the Massachusetts Institute of Technology.

The Boston-Cambridge area was a focal point for the environmental movement, including the first Earth Day on April 22, 1970. The passage of the Clean Air Act of 1970 and the creation of the EPA by President Richard Nixon were widely celebrated in Boston and Cambridge as transformative events. On completing his master of science degree, Doug was inspired to use his training in meteorology and worked as a scientist for the new EPA in its New England Regional Office.

In the mid-1970s, Doug pursued additional training at the Harvard School of Public Health, where he graduated with a doctoral degree in environmental health in 1979. By 1989, Doug was well known for his studies of air pollution exposures and their health effects. He had been working with a team of innovative scientists since 1974 on the groundbreaking Harvard Six Cities study. In March 1989, two months before the publication of Arden's research regarding the steel mill, Doug and colleagues published a research article reporting that children exposed to air pollution had a notably greater risk of respiratory illness.[3]

Though Doug and Arden worked in very different fields, their research led them to similar conclusions about the connections between air pollution and children's health. So, how did they get together several months after that November 1989 morning?

Arden and Doug were initially connected by Janet Raloff, a reporter with *Science News*, who wrote an article about their two recently published research papers.[4] When interviewed by Raloff for the article, Arden learned of the study by Doug and his colleagues on the effects of long-term air pollution exposure and children's respiratory disease. When Raloff asked Doug about Arden's research, Doug enthusiastically responded that he thought it was "a landmark study" based on its unique and valuable natural experimental design.

After reading Doug's response to his study in Raloff's article, Arden phoned Doug. He wanted to visit and hopefully get help and advice on how to understand and interpret his research in Utah Valley. He also was dealing with hostile controversy and criticism in the local press for implicating a significant economic resource in the community as contributing to children's respiratory disease. Arden had already instigated a follow-up study, but he was searching for suggestions on how to test further the hypothesis that local air pollution was affecting the health of the Utah Valley population.

As it happened, Doug and colleagues were also preparing a new study—an extensive study of acute changes in lung function of asthmatic children. Arden and Doug discussed a potential spin-off study using panels of schoolchildren living near Geneva Steel. Doug invited Arden to give a seminar at Harvard and to meet with the full team of Harvard investigators. That visit in November 1989 began a collaboration on panel studies of children living near the steel mill. It led to Arden's 1992/93 sabbatical to Harvard, and it initiated a friendship and academic partnership that continues, even to the coauthoring of this book. Figure 2 shows photos of

Figure 2
Photos of authors: Arden Pope (left), Doug Dockery (right). Top: Utah Valley in the background, circa 1995. Bottom: Harvard University symposium in honor of Doug, October 2022. *Sources:* Top: Authors; bottom: Steve Gilbert, Studioflex Productions.

the authors taken while conducting early studies in Utah Valley and later while working on this book.

Writing this book has brought back many memories. But this book is not a memoir. In this book, we attempt to tell the exciting and crucial story of the science of air pollution and human health. We tell this story from the perspective of two academics who conducted research on air pollution and human health for almost four decades and participated in many key scientific findings and public policy controversies.

Throughout our research journey, we met and established collaborations and friendships with many incredible researchers and others engaged in efforts to have healthier air. They are key characters in this book. They include medical researchers, epidemiologists, engineers, meteorologists, toxicologists, chemists, statisticians, econometricians, and other researchers—across a vast array of scientific disciplines. Furthermore, this book begins with stories of children seriously impacted by air pollution and the efforts of their parents to understand and help chart a course for cleaner air. The stories of the contributions of all these remarkable people bring this book to life.

At one point, Doug observed that when we talk about our colleagues' research, we commonly refer to them as skilled, excellent, well-respected, innovative, highly productive, and other favorable terms. How fortunate we have been to work with and learn from such scholars and professionals. Although we wrote this book from our own perspective, we are intensely aware that the story of air pollution and health science is a story that involves many skilled, excellent, innovative, and productive people.

This is a story that involves more than just polluted, unhealthy air but also controversy and sometimes toxic debate. There has been rancorous and openly hostile opposition to scientific findings and sometimes even to individual scientists. The scrutiny over air pollution research has been national and even international in scope. Our efforts to address the various controversies primarily includes trying to conduct high-quality research that addresses legitimate concerns and uncertainties. We applied new and innovative approaches to study air pollution and health. We tried to synthesize and understand that information. For example, we collaborated on research efforts that included natural experiments, panel studies of children's lung health. We conducted studies of short-term and long-term air pollution exposure and mortality. We conducted studies on air pollution's impact on aging and life expectancies, and explored how air pollution can lead to disease and death. We coauthored systematic and critical reviews of the scientific evidence of the health effects of air pollution.

Some of our efforts were more successful than others, and there have been false leads and disappointments. We value study designs and analyses that are elegant, straightforward, and understandable. This journey had no road map, and we learned as we went. We always tried to be critically wary of observed results, particularly our own. We valued thoughtful criticism, but when subjected to spurious critiques, we tried to follow the data and pursue the evidence wherever it led.

One of the biggest challenges of writing this book is that the story of air pollution and health science, along with the accompanying controversy, cannot be easily told in one simple, overarching, classic narrative arc. Advances in research were not always systematic and tended to be challenged. This scrutiny typically motivated further research. The chapters of this book are organized to present the advances in scientific evidence and to address the accompanying waves of scrutiny and primary controversies in an orderly and integrated way.

We traveled around the world to work with colleagues on various research projects and to attend and participate in conferences, meetings, and other events. We served on research oversight committees and advisory panels. We attempted to inform but not to dictate public policy. Writing this book prompted us to contemplate the decades of effort we have put into this research. Although there have been substantial challenges, it has been an extraordinary privilege to do this work. We can think of no better career.

Finally, we know that the work is not complete. Much remains to be learned regarding the effects of various air pollutants on human health and welfare. There is a fresh generation of excellent air pollution researchers, and new studies are appearing at a rapid pace.

More importantly, there remains much to be done to find reliable, cost-effective, politically feasible approaches and strategies to reduce and control air pollution. Finding ways to provide healthy air will greatly benefit human health and welfare. Furthermore, understanding and addressing the inextricable links between various air pollutants, human health, and the climate is critical to the health and quality of our earth's environment.

This is the story of the air that we all breathe, and what we must do together to keep it breathable and healthy. We ask you to keep this in mind as you read the chapters ahead. Because it's your story, too.

1

WHAT IS HEALTHY AIR, AND WHY DOES IT MATTER?

We have studied the health effects of air pollution around the world. We have learned that, although these effects are far-reaching and seemingly abstract, they are deeply personal to those affected and their families and loved ones. Let's take a close-up look.

Ella Kissi-Debrah was a popular, fun-loving little girl who engaged in sports, dancing, singing, and acting. She was in the top 10 percent of students in her school. The Royal Air Force Aerobatic Team, called the Red Arrows, inspired her. She dreamed of being an air ambulance doctor who helped to rescue people. Tragically, Ella's health, her ability to participate in activities that she loved, and even her dreams were thwarted by air pollution.[1]

Ella lived in southeast London just a few dozen yards from Britain's most notoriously congested roadway. She was chronically exposed to fine particulate matter—including microscopic bits of particulate matter produced by burning—as well as nitrogen dioxide air pollution from cars and trucks stuck in traffic outside her home and as she walked to school along this roadway. At age six, Ella developed a chest infection, a persistent aggravating cough, and alarming seizures. Her medical doctors were puzzled, and they dutifully tested her for various diseases, including cystic fibrosis and epilepsy. Ella was eventually diagnosed with severe asthma.

Ella was first hospitalized for a severe coughing fit at age seven. The following twenty-eight months were worse than anyone could imagine. She was admitted to the hospital twenty-seven times. Rosamund, Ella's mother, was coached to treat her attacks, but no one could tell her what was causing them. Then, in February 2013, at age nine, Ella suffered a fatal attack. Her death certificate reported that she died of acute respiratory failure.

Stephen Holgate, a respiratory physician specializing in asthma and air pollution, heard about Ella and agreed to review her case. He agreed with the finding that Ella had severe asthma. More importantly, he found that her hospital admissions coincided with periods of high air pollution. Ella's anguished mother learned that pollution contributed to her daughter's disease and death. She lamented that if she had only known about air pollution and its impact on her daughter's health, she would have moved to a less polluted place to protect Ella.[2]

Rosamund remains deeply pained by the loss of her young daughter, yet she has honored Ella by passionately advocating for cleaner air (figure 1.1). After learning about the likelihood that air pollution contributed to Ella's respiratory disease, Rosamund fought to reopen the inquiry into her daughter's death. She believed that cleaner air could reduce the destructive health effects of air pollution, sparing other families the same fate. As a result, a formal "Report to Prevent Future Deaths" stated that air pollution significantly contributed to Ella's disease and death.[3] Rosamund also established a foundation, "Clean Air for All," in honor of her daughter.[4]

London is just one of many places where air pollution threatens children's health. In Utah Valley, in the Rocky Mountain Wasatch Front of north-central Utah, another fun-loving little girl and her family struggled with the alarming health effects of air pollution. In the 1980s and 1990s, people living in Utah Valley commonly experienced air pollution episodes. Kristina was especially vulnerable to air pollution exposure. Arden and Doug visited with Kristina's parents, Kimberly and Ned Warner, in November 2022.

After Kristina was born, both parents quickly realized that air pollution posed a serious threat to their daughter's health. During episodes of elevated air pollution, Kristina would become seriously ill with alarming

Figure 1.1
Mayor of London Sadiq Khan with Ella's mother, Rosamund Adoo-Kissi-Debrah, in southeast London, on the first day of the expansion of the ultra-low emission zone, August 29, 2023. *Source:* PA Images/Alamy Stock Photo.

breathing problems. Her mother remembered, "It seemed as if she was suffocating, especially if she was outdoors during an air pollution episode." Kristina seemed to breathe better in a sitting position, but her parents often anxiously worried if she would be able to take another breath. Like Ella, Kristina had alarming and harmful seizures.

The physicians who treated Kristina told her parents that air pollution was an important contributor to Kristina's breathing problems. Ned and Kimberly were able to help Kristina avoid the air pollution in the valley. They installed a high-quality air filtration system in their home and kept Kristina indoors whenever air pollution increased. They even purchased another home near a local ski resort, up and out of the more polluted valley, and used it as a place to further escape the air pollution.

Kimberly joined with other parents concerned about the health impacts of air pollution and was a cofounder of the Utah County Clean Air Coalition. They began a local fight for healthy air. This effort included the production and distribution of an eight-page brochure that provided information about local air pollution and included unidentified photos of local citizens who were highly sensitive to air pollution (figure 1.2).

Utah Valley's air pollution was less dense than the historically severe air pollution episodes (see chapter 2). Nevertheless, the parents in the

Figure 1.2
A cover photo from the Utah County Clean Air Coalition brochure. Air pollution trapped in Utah Valley in the background with an inserted photo of three individuals who were highly sensitive to local air pollution. *Source:* Adapted from a color cover photo from Utah County Clean Air Coalition brochure, February 1991.

coalition were especially concerned about the largest and most obvious local source of air pollution in the valley, a large World War II–era steel mill.

Owners and operators of the steel mill vociferously denied that their pollution could harm children's health. Kristina's parents knew otherwise.[5] The Warners' commitment to air quality in the valley was bolstered by local studies showing that episodes of elevated air pollution were associated with increased respiratory symptoms, reduced lung function, and hospitalizations among children in Utah Valley.[6]

Kristina is now an adult and still lives in Utah Valley. The steel mill in the valley has closed, and air quality has improved. Although Kristina continues to have health challenges, her respiratory health has also improved. It is impossible to tell how her health is still affected by any long-lasting effects of air pollution.

Could Ella's outcome have been different if her mother had the resources, including the information, that Kristina's parents had? Possibly so. Health equity and environmental justice are critical elements in the story of air pollution.

Air pollution's impacts on human health are complicated and differ substantially for various individuals and life stages. Yet some patterns are linked to socioeconomic status, race, and other characteristics (see chapter 6). The stories of Ella and Kristina underscore the importance of having the means and ability to advocate for clean air, avoid exposure to air pollution, and live in less polluted places. Air pollution affects individuals and sections of populations differently based on levels of exposure, susceptibilities, socioeconomic status, abilities to avoid or minimize exposure, and even differences in access to adequate and informed medical care.

While air pollution contributes to disease and even death in children, most deaths occur in adults. Air pollution's contributions to cancer, respiratory, and cardiovascular disease are not easily distinguished from other factors that affect these chronic diseases. The cumulative, insidious effects of long-term repeated air pollution exposures on cardiopulmonary disease can accelerate disease progression. These effects may be disguised as part of the natural aging process.

Contributions of air pollution to disease are rarely indicated on death certificates. Individual accounts or anecdotes of people affected by air pollution, including Ella and Kristina, help illuminate the critical importance of addressing the health effects of exposure to air pollution. Additionally, decades of extensive and rigorous scientific studies provide us with an understanding of the health effects of air pollution. One important challenge of this book is to accurately summarize this extensive scientific literature, making clear what the stakes are, so that we can make better decisions now and in the future. These decisions include not only public policy choices, but also personal and family choices regarding our contributions to pollution and where to live, work, study, exercise, and play.

OBJECTIVES OF THIS BOOK

This book has three essential and interrelated objectives. The first is *to present scientific evidence regarding the human health effects of air pollution.* You will learn not just what scientists know about air pollution but how they know it. Scientific research on this topic has grown dramatically since the mid-1980s. This research is compelling and fascinating. It provides evidence that air pollution is one of the most significant modifiable risk factors for global disease and death. Air pollution contributes to an estimated 4.2 million deaths annually and is estimated as one of the top five most prominent risk factors contributing to the global disease burden.[7]

The death of Ella Kissi-Debrah was a tragedy. But 4.2 million is a difficult-to-grasp statistic. What is the evidence that air pollution inflicts such havoc on human health, contributing to disease and millions of premature deaths? Where does this evidence come from?

A challenge with presenting scientific evidence of human health effects of air pollution is that the primary evidence has been published in many academic and scientific journals over decades. These include journals of medicine, epidemiology, toxicology, public health, biostatistics, environmental science, and more. This book is not a comprehensive critical review of this extensive literature. Such reviews are available, have been published in scientific literature, and will be referenced when relevant.

The intent of this book is to present this scientific evidence in a way that is broadly accessible and understandable to the larger community.

The second objective of this book is *to address key scientific and public policy controversies*. These controversies include efforts to use scientific evidence to establish air quality standards and control air pollution. There have been substantial attempts to cast doubt on scientific studies indicating that exposure to air pollution contributes to an increased risk of respiratory, cardiovascular, and cancer disease and death. The science of air pollution and health has been challenging and has involved a diverse group of people with different perspectives, much like other sciences. And much like other sciences, there have been major disputes, uncertainties, and disagreements regarding how to conduct studies and interpret and use the evidence.

But some disagreements have gone beyond genuine scientific disputes. Some deny the health costs of air pollution in attempts to avoid or delay the expense of the air pollution abatement that would protect public health.

How much of this controversy involves legitimate scientific uncertainty? How much of this controversy involves the manufacturing of doubt to metaphorically "muddy the waters" or "create a smokescreen"? Books like David Michaels's disturbing *Doubt Is Their Product: How Industry's Assault on Science Threatens Your Health*[8] or Naomi Oreskes and Erik Conway's equally troubling book *Merchants of Doubt: How a Handful of Scientists Obscured the Truth on Issues from Tobacco Smoke to Global Warming*[9] cover this question. Concerted attempts to cast doubt on air pollution and health science are often efforts to resist public policy initiatives to reduce air pollution.

This book's third objective is *to tell an important and compelling story* of science history. Historical and contemporary efforts to understand the health effects of air pollution have woven a fascinating and sometimes contentious tale. This book tells part of this story; the entire story could not be told in any single book.

This book tells the story of discovery and controversy from the authors' perspective, the perspective of two researchers who have contributed to research on air pollution and human health for almost four decades. The authors have also been significantly involved in critical scientific and public policy controversies. They are intensely aware of, respect, and admire the many skilled and dedicated air pollution researchers, including those with different perspectives. Many are their friends and colleagues.

The story of air pollution and human health could begin with pre-historic humans affected by cooking or heating fires in caves or huts.[10] Alternatively, it could start with respiratory disease from air pollution experienced during medieval London.[11] For this book, however, the story begins with the emergence of modern capitalistic, free-market econo-mies, the Industrial Revolution, and remarkable industrial growth—along with the emergence of areas with extremely high levels of air pollution.

Several severe air pollution episodes demonstrated substantial disease and death, motivating major public policy efforts to control air pollution. These included historic Clean Air Act legislation and the creation of the U.S. Environmental Protection Agency in 1970.

The story also includes subsequent decades of research exploring the health effects of air pollution, with extensive academic literature pub-lished in leading scientific journals. Another intriguing aspect is this story includes accusations of junk science, "secret science," scientific mis-conduct, and investigative reanalyzing. It further includes debates over causation, biological plausibility, environmental justice, economic feasi-bility, public health, and climate policy.

We will be telling this story in chapters focusing on eleven crucial sci-entific and public policy questions that have been controversial. You'll read the scientific evidence, dive into the scientific and public policy con-troversies, and hopefully come away with a deeper understanding of the dangers of air pollution and the opportunities to make changes to protect the air we breathe.

WHAT IS UNHEALTHY AIR?

Clean, pure air consists primarily of nitrogen dioxide (78 percent), oxygen (21 percent), argon (0.9 percent), trace amounts of other gases, and water vapor. Clean, pure air is well-suited to sustain human life and other life forms, including plants and other animals. Unhealthy air is polluted or contaminated with gaseous and particulate matter pollution, including the combination of gases such as nitrogen dioxide and the microscopic particles from cars and trucks, power plants, industry, and burning. As discussed in chapter 2, air pollution episodes can result in air so thick and so dense that sunlight is blocked, visibility is minimal, and humans and animals struggle to breathe. However, even the moderately polluted air we regularly find in modern cities can harm human health and welfare.

Figure 1.3 provides a simple stylized illustration of common air pollutants generated from combustion processes, such as vehicles burning gasoline and diesel fuels; coal and wood burning; industrial processes such as smelters, steel mills, and cement plants; wildfires; and more.

Fires need not be powered by wood or coal or come from industrial smokestacks to be unhealthy if smoke builds up. In parts of the developing world, all sorts of things are burned in heating and cooking fires. Jon Krakauer, a mountain climber and author of *Into Thin Air*, a book about his ascent of Everest, described an encounter with badly polluted air on his journey. While trekking to base camp on Mount Everest, he and his companions stopped at Lobuje, Nepal, a small village at the edge of the iconic Khumbu Glacier and one of the last overnight stops on the trek to base camp. They spent the night in a primitive, cramped, and filthy lodge. The night was cold, and the lodge was heated only by a small iron stove that burned dried yak dung. Krakauer writes:

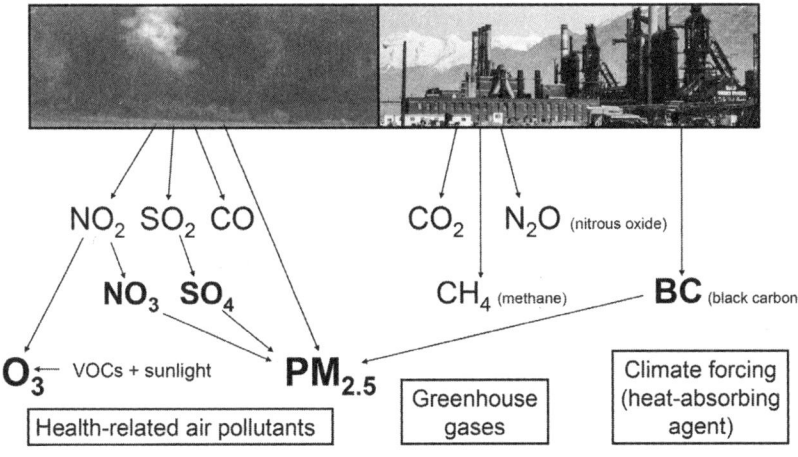

Figure 1.3
Common air pollutants from combustion and industrial processes. These include vehicle use of gasoline and diesel fuels, coal and wood burning, and industrial processes such as smelters, steel mills, and cement plants; wildfires; and more.

The lodge filled with dense, acrid smoke, as if the exhaust from a diesel bus were being piped directly into the room. Twice during the night, coughing uncontrollably, I had to flee outside for air. By morning my eyes were burning and bloodshot, my nostrils were clogged with black soot, and I'd developed a dry, persistent hack that would stay with me until the end of the expedition.[12]

What's in the unhealthy air that causes this kind of misery?

The title of this book, *Particles of Truth*, alludes to the discovery of the impact of particles in the air, or particulate matter (PM) air pollution, on human health. PM air pollution consists of particles of different sizes and from different sources. Because PM is carried in the air and deposited in the lungs, a relevant measure of particle size is aerodynamic diameter, usually measured in micrometers. Relatively large particles do not remain suspended in the air for very long and are not easily inhaled. Particles smaller than 10 micrometers, or PM_{10}, are more easily inhaled into the lungs and are called inhalable particles. Particles smaller than 2.5 micrometers, or $PM_{2.5}$, are called fine particulate matter. These particles are tiny. The diameter of a human hair averages about 70 micrometers; $PM_{2.5}$ particles range in size from less than 0.1 to 2.5 micrometers. They would be just tiny specks on the end of a human hair and can be seen only through a microscope.

Fine particulate matter ($PM_{2.5}$) is the air pollutant most strongly associated with widespread adverse human health effects. Unlike relatively larger particles, $PM_{2.5}$ can be breathed deeply into the lungs, and the smallest $PM_{2.5}$ particles can even pass through lung tissue and enter the bloodstream. These fine particles are made up of a chemically complex mixture of combustion-related byproducts, including soot, nitrate, sulfate, and other primary and secondary combustion particles.[13] Further, these fine particles remain suspended in the air for long periods and can disperse over relatively large areas, exposing large populations.

Other important air pollutants, including the various gaseous pollutants, nitrogen dioxide (NO_2), sulfur dioxide (SO_2), carbon monoxide (CO), and volatile organic compounds (VOCs), contribute to the mixture of unhealthy air. Interestingly, some gaseous pollutants can contribute to particulate matter pollution when they interact in the lower atmosphere. For example, the gaseous pollutants NO_2 and SO_2 help form nitrate particles (NO_3) and sulfate particles (SO_4)—which are important constituents of overall $PM_{2.5}$. NO_2 and VOCs, along with sunlight, also contribute to the formation of ground-level ozone (O_3), which further adds to the health burden of air pollution. Carbon dioxide (CO_2), nitrous oxide (N_2O), and methane (CH_4) are important greenhouse gases. Black carbon (BC), another important constituent of overall $PM_{2.5}$ particulate matter, not only threatens human health but also contributes to climate change.

$PM_{2.5}$ is a complicated mix of fine particles partly because it is generated by so many different sources, including large factories, power plants, gasoline and diesel vehicles, or even small fires like the one in Nepal that Jon Krakauer described. $PM_{2.5}$ was a primary pollutant that contributed to sickening Ella and Kristina in London and Utah Valley.

CONCLUSION

"What is unhealthy air, and why does it matter?"

"Unhealthy air" is laden with particles or particulate matter, some so small they can be breathed deeply into the lungs and, for some particles, transported into the bloodstream. High concentrations of fine particles, $PM_{2.5}$, are especially insidious. The higher the concentration of these particles, the less healthy the air. Research has revealed that $PM_{2.5}$ inhaled into the lungs can initiate and exacerbate many diseases, including respiratory, cardiovascular, and other conditions.[14] Even brief exposure to high doses of $PM_{2.5}$ can have lingering effects.

2

DOES AIR POLLUTION SMELL LIKE MONEY OR DISEASE?

Starting in the mid- to late 1700s, Great Britain saw conspicuous advancements in industrial innovation and production. These advancements spread to continental Europe and the United States. The rapid growth in industrial innovation and production was so substantial that it was deemed a revolution—the Industrial Revolution (approximately 1750–1914). Extraordinary technological innovations occurred in industries such as textiles, iron and steel, steam power, machinery, chemicals, mining, transportation, and more.

The Industrial Revolution was facilitated by stable political systems and governments that largely adhered to the doctrine of laissez-faire (see box 2.1), leaving businesses on their own to compete freely with one another. Laissez-faire made possible largely unfettered free-market industrial activity, with little restraint by the government, even for public health and safety. Such political systems and governments played a role in the emergence of modern capitalistic economies.

There have been, and continue to be, massive human benefits from innovations, technologies, and production from a myriad of industrial enterprises competing in free-market economies. But the Industrial Revolution also birthed stunning growth in pollution. In heavily industrial cities, substantial increases in air pollution were often distressing and oppressive (figure 2.1). Public and scientific concerns about air, water,

---

Box 2.1
The Doctrine of Laissez-Faire

Doctrines or core belief systems profoundly influence human and societal behavior and welfare. One such doctrine, the doctrine of laissez-faire, has had critical historical, political, and economic impacts. It also has had substantial environmental impacts—including air pollution.

Jean-Baptiste Colbert (1619–1683) was born into a French family of merchant businessmen. He served as a prominent minister in France, dealing with finances and commerce under King Louis XIV. As part of his economic reform efforts, Colbert purportedly asked a group of businessmen and industrialists how the government could help business and commerce. Their answer, in French, was "Laissez-faire," loosely translated as "leave us alone" or "let us be."

The doctrine of laissez-faire was developed by classical economists, most notably by Adam Smith (1723–1790) in his influential 1776 book, *The Wealth of Nations*. Applied to economic policy, this doctrine prescribes minimal or at least limited government interference in the workings of free markets. It extols the benefits of unfettered, free, economic enterprise.

This doctrine should not be misunderstood as suggesting a carefree, laid-back attitude. It does not offer calming, comforting words of wisdom as suggested in the famous Beatles ballad, "Let It Be." In the context of economics, the doctrine of laissez-faire indicates an attitude that industry and business should be largely left alone and allowed to compete actively, aggressively, and vigorously in free, open, capitalistic markets. Rigorous open-market competition will result in business failure as well as success. Successful businesses will include those that produce most efficiently, are highly enterprising, are innovative, and may be able to specialize effectively—those that can survive in open competitive markets.

---

and land pollution grew. Advocates during the Industrial Revolution attempted to mitigate air pollution through public policy. Peter Brimblecombe's book, *The Big Smoke*, documents some of these early efforts. Yet these attempts to mitigate air pollution during the Industrial Revolution and for decades following were largely ineffectual.

Financial objections were barriers to air pollution control strategies. Industrial polluters appealed to various versions of the doctrine of laissez-faire (box 2.1) in making their cases. Furthermore, the public did not adequately understand the potential health effects of pollution.

Figure 2.1
Smells like money to me. Widnes, England, late 1880s. *Source:* D. W. F. Hardie, *A History of the Chemical Industry in Widnes* (London: Imperial Chemical Industries Limited, 1950), https://commons.wikimedia.org/wiki/File:Widnes_Smoke.jpg.

Although there was minimal scientific evidence, some forward-thinking individuals had concerns that air pollution could contribute to disease. Others, perhaps those with faith in the doctrine of laissez-faire, viewed air pollution as a necessary byproduct of the industrial development spawned by free economic enterprise. When confronted with the foul odors of polluted air, they sardonically responded, "Smells like money to me."

But does air pollution have adverse health effects? And how would scientists know if it did? Let's look at some experiments.

NATURAL EXPERIMENTS: AIR POLLUTION AND HUMAN HEALTH

Imagine an experiment to study the human health effects of air pollution. If it were a standard controlled experiment, you would have some subjects kept clear of any air pollution—this would be the control group—and other subjects who would be exposed to a high level of air pollution. After some time, perhaps months or years, you'd compare the health of members in each group to see if there was a substantial difference between them.

However, this approach is impractical and unethical. Clearly, you wouldn't deliberately expose people to dangerously elevated levels of air pollution. Experiments like this are off the table for moral and ethical reasons, so researchers must find alternative approaches.

One thing researchers can do is look for situations in which people have been exposed to elevated air pollution without any involvement or experimental control by scientists. Such situations are called "natural experiments" to distinguish them from deliberate experiments.

Some of the most straightforward and compelling studies of the health effects of air pollution are facilitated by a natural weather phenomenon called a *temperature inversion* (also called a thermal inversion). Under typical atmospheric conditions, the air is warmer nearer the ground and colder at higher altitudes. The warmer, lighter air at the surface tends to rise, and the colder, denser air above tends to sink. This overturning or mixing of the atmosphere dilutes air pollutants emitted near the ground. A temperature inversion occurs when this normal situation is "inverted," trapping cold air near the ground under a layer of warmer air. These conditions commonly occur with low winds and a stationary high atmospheric pressure system (sometimes called an anticyclone). During a temperature inversion, air pollutants released near the ground are not mixed upward, so pollution concentrations build up over time.

Temperature inversions can be especially problematic in valleys or basins. The warm layer serves as a lid, trapping the colder air on the bottom of the valley floor, where people live and where pollution is usually produced. Pollutants can accumulate over days, resulting in episodes of dramatically elevated air pollution.

Major temperature inversion events can serve as informative natural experiments. Although there is no experimental control—no observed group kept apart from the air pollution—a temperature inversion provides an opportunity to observe adverse health effects in a fixed population before, during, and after the air pollution episode. The valley is essentially a natural human exposure chamber. The following are some examples of such natural experiments.

MEUSE VALLEY, BELGIUM (DECEMBER 1–5, 1930)

In 1930, an unusually strong temperature inversion settled over the Meuse Valley in Belgium. The Meuse Valley follows a twelve-mile stretch of the Meuse River from Liege to Huy, Belgium. While it is a scenic river valley today, it was heavily industrialized in 1930, with approximately twenty-seven factories, including steelworks, zinc smelters, glass manufacturers, and explosives plants. Air pollution was emitted from these industrial sites, with additional emissions from domestic coal burning.

During five days in early December, an inversion layer settled about a hundred yards above the valley floor, trapping the air pollution within the hills on either side. Although nobody monitored the elevated concentrations of air pollutants at the time, the health consequences of the pollution were evident. By December 5, there were sixty-three deaths, about ten times the expected number.

Engis, a community near the valley's center with a population of about 3,500, had the most deaths. Hundreds in the community experienced severe respiratory symptoms, including throat irritation, chest pain, coughing fits, and asthma-like symptoms. Cattle also experienced respiratory problems, requiring many to be slaughtered. The respiratory illnesses in the population improved markedly after the thermal inversion passed and the air pollution dissipated.

The 1930 Meuse Valley air pollution episode harmed human health. But did it impact air pollution policies? In May 1931, a report (in French) on the event was presented to the Royal Academy of Medicine of Belgium. A short version published in English by Jean Firket noted that "public authorities were anxious to know the causes of this catastrophe."[1] A commission was appointed to evaluate additional legislation regarding industrial air pollution. A book about the Meuse Valley disaster, which included the original report, recommended "the British approach of monitoring air pollution." It also stated that "little was done since air pollution was considered an unavoidable consequence of prosperity."[2]

An unavoidable consequence of prosperity? Is it true that economic prosperity cannot be achieved without seriously harmful air pollution?

Interestingly, in the English version of the report, Firket included a clear warning. If an event like the one in Meuse Valley occurred in the much larger population of London, there could be up to 3,200 sudden deaths.[3]

DONORA, PENNSYLVANIA (OCTOBER 26–31, 1948)

In 1948, Donora, Pennsylvania (including the small neighboring town of Webster), was a community of about 14,000 people located twenty-five miles south of Pittsburgh. The community is in a valley on a bend of the Monongahela River, hemmed in by surrounding hills and bluffs. In earlier times, this river bend and its immediate surroundings were part of a pastoral, even bucolic, picturesque farming community.[4] By 1948, however, it was a gritty industrial area. Its workers were employed locally in steel, wire, or zinc plants. These were large mills, complete with numerous smokestacks constantly emitting mingling plumes of black, red, or yellow smoke (see figure 2.2). The inhabitants of Donora were familiar with air pollution. The town's confined location allowed pollution to linger. Yet Donora residents were not prepared for what would happen during the last week of October 1948.

Beginning on October 26 or 27, a temperature inversion began trapping pollution into Donora's hemmed-in bend of the Monongahela. The highly polluted air often was referred to as fog. But it was more sinister than fog. Thick, stagnant unhealthy air immersed the community. Visibility declined dramatically, making it difficult to get around town or even visit the ill or dying. The smog smelled bad, reeking of the coal-burning and smelting operations, only worse than usual. At noon, it was nearly as dark as night.

During the episode, as air pollution intensified, Donora residents started becoming ill, many seriously ill. They reported irritating and increasingly severe health problems: sore throats, headaches, coughing fits, and trouble breathing. Local doctors were inundated at their offices and over the phone with requests for medical help. But there was another problem. Doctors were becoming ill themselves. One of them, Dr. Edward Roth, concluded they were dealing with something serious. He was worried but not bewildered. To him, the sharp increase in illnesses was not

Figure 2.2
Several smokestacks and a factory in Donora, Pennsylvania. *Source:* NOAA National Ocean Service, "A Brief History of Pollution," https://oceanservice.noaa.gov/education/tutorial_pollution/02history.html.

a mystery. He indicated that "it was obvious—all the symptoms pointed to it—that the fog and smoke were to blame."[5] Dr. Roth's own symptoms included chest tightness, uncontrollable coughing, choking, and feeling terribly sick. He gave himself an injection of adrenaline, began to feel better, and continued attending to patients who were sick and miserable. At one point, Dr. Roth could hear the town's annual Halloween parade. He was incredulous. "People were cheering and yelling, and the bands were playing. I could hardly believe my ears. It just didn't seem possible."[6]

A few days into the air pollution episode, the eight doctors in town were overwhelmed with emergency calls and patients in their offices stricken by the polluted air. They tried to persevere, although they also were coughing, choking, and having difficulty breathing, with chest tightness and even heart problems. The doctors could not keep up with the urgent needs of those acutely affected by the pollution. The thick, polluted air reduced visibility so much that it was difficult to make house calls. Many of the more seriously afflicted Donorans, desperate and unable to reach a doctor, began calling the Donora Fire Department, begging for assistance, often requesting oxygen from the firefighter's inhalators.

The Donora Fire Department had about thirty volunteers but only two full-time firefighters: Chief John Volk and his assistant, Russell Davis. Volk and Davis were concerned about whether they should be treating the sick, but they did anyway, call after call. They recruited Bill Schempp, a volunteer firefighter, to bring his compressed oxygen tank. They were afraid that they would run out of oxygen—and they did. Over the next few days, they borrowed oxygen from McKeesport, Monessen, Mononga-hela, Charleroi, and other neighboring communities. They worked fast and hard for days. They had to decide, with limited training, how much oxygen they could or should give one ailing person before hurrying to the next sufferer pleading for air.[7]

On the third day of the episode alone, fifteen people died. An additional five deaths were attributed to the episode. These twenty deaths were about ten times the normal number, resulting in a local shortage of caskets. How many more would have died had it not been for the valiant efforts of many citizens of Donora, including doctors, nurses, and firefighters? Nearly 6,000 persons (or about 43 percent of the population) experienced illness associated with the smog episode.[8] After four days, relief came with changes in weather, a breakup of the thermal inversion, a good rainstorm, and much cleaner air.

It is challenging to know the full impact of the Donora smog episode. A follow-up study demonstrated that those who became ill during the episode had higher mortality and prevalence of illness ten years later.[9] The episode, coming about eighteen years after Meuse Valley, served as a second example of the severe health effects of extreme air pollution. It sparked further alarm and debate about what public policy measures should address air pollution. Lynne Page Snyder later argued that "within the public health community and among federal policymakers, the Donora smog catalyzed a new approach to air pollution."[10]

She may have been right, but at least one more major catalytic jolt was coming.

LONDON SMOG (DECEMBER 5–9, 1952)

London lies on the low ground along the River Thames, surrounded by hills that define the Thames river basin. Predominant southwest winds typically clear air pollution out of this basin-like topography. However, during temperature inversions, London is susceptible to still, stagnant, foggy air. Air pollution is easily trapped in this natural basin.

As in Meuse Valley and Donora, London air was contaminated with pollution from multiple sources, including emissions from industrial smokestacks, chimneys from coal-burning homes, and vehicle tailpipes. London's air pollution in the early 1900s is illustrated by the artist Claude Monet in many paintings of the city that depicted the Waterloo Bridge and the Houses of Parliament. The dirty, dense air in the smoke-contaminated fog was descriptively referred to as *smog*, a combination of smoke and fog.

Residents of London were accustomed to frequent smog episodes, referring to them as "pea soupers." Kate Winkler Dawson notes, "It was obnoxious, the dirty air, but most Londoners accepted that it was their penalty for living in the world's most urbanized and industrialized city."[11]

For Londoners still trying to recover from the devastation of World War II, air pollution was not necessarily viewed as an "unavoidable consequence of prosperity," as was suggested following the Meuse Valley episode in 1930. But London relied on dirty coal for domestic heating, and dense industrial activity was a means of moving forward, surviving the aftermath of war, and recovering prosperity.

On December 5, 1952, an exceptionally strong and persistent temperature inversion (including high pressure with cold temperatures and little wind) settled over London. London's most disastrous air pollution episode commenced. The thick, brownish-yellowish, grimy smog, with a choking smell, filled the air and shrouded the entire city. The smog was so thick that visibility dramatically deteriorated. At its worst, people could not see for more than a few yards, even in the daytime. As visibility deteriorated, transportation around the city was severely compromised. Sporting events were canceled. People got lost even while traveling on foot. There are many stories documenting the difficulties and challenges that the smog caused the residents of London.[12]

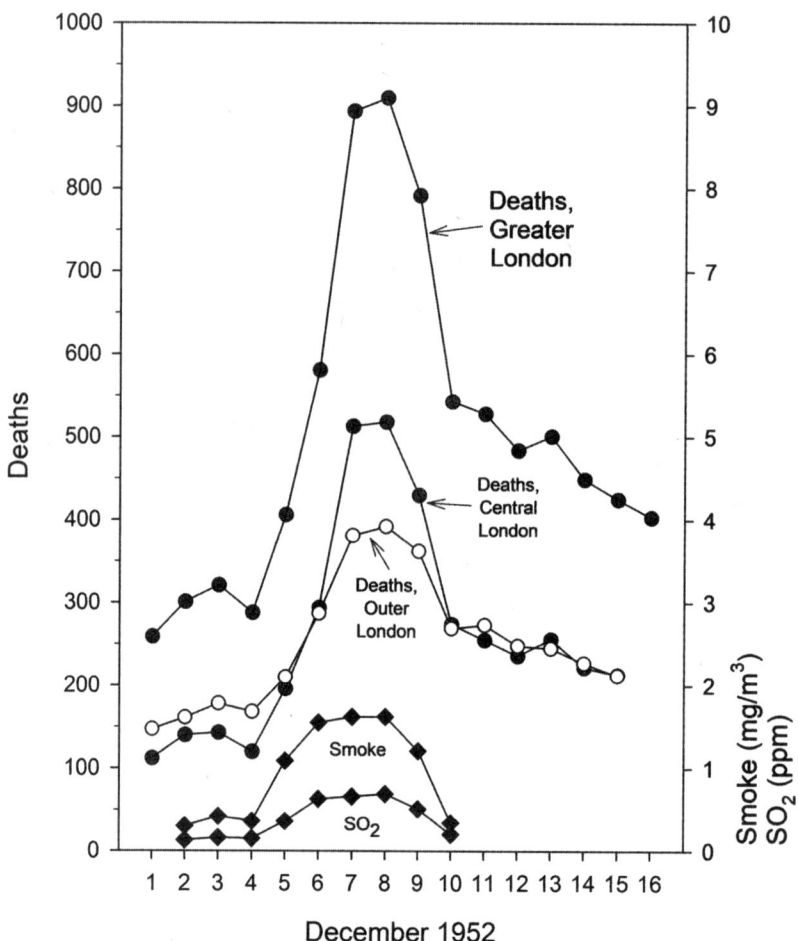

Figure 2.3

Daily deaths and air pollution, 1952 London smog episode. *Source:* Illustration created using data from W. P. D. Logan, "Mortality in London Fog Incident, 1952," *Lancet* 1, no. 6755 (1953): 336–338, and U.K. Ministry of Health, *Mortality and Morbidity during the London Fog of December 1952*, Reports on Public Health and Medical Subjects No. 95 (London: Her Majesty's Stationery Office, 1954).

The severe smog episode lasted for five days. And it was inescapable. It penetrated businesses. It penetrated theaters. It penetrated homes. It penetrated lungs. At the Smithfield Cattle Show in London, prized livestock suffered from breathing difficulties. Some of the cattle died, and others had to be slaughtered. David Bates was a young medical doctor who worked at London's St. Bartholomew's Hospital. Trying to understand how his human patients died from exposure to air pollution, he examined lung sections of cattle that died during the event to see what killed these animals. He concluded that the animals died of "a very acute inflammation in the small airways of the lungs . . . an inflammation in response to the fog they had been breathing."[13] David became a pioneering air pollution epidemiologist, primarily motivated by his experiences during the London smog episode.

For humans in London, this episode of unhealthy, polluted air helped develop and exacerbate disease, especially respiratory and cardiovascular conditions. People suffered, and many died. Figure 2.3 presents the daily deaths in London shortly before, during, and after the episode for the Greater London area, Central London (London Administrative County), and Outer London (the outer ring by county areas). The episode-related increase in deaths in Greater London is evident. The relative increase in deaths was larger for Central London (where pollution sources were densest) than for Outer London. The initial estimate of the number of excess deaths attributed to this pollution episode was approximately 4,000.[14] Later analyses that accounted for more persistent effects of exposure estimated as many as 12,000 excess deaths attributable to this episode.[15]

PUBLIC POLICY RESPONSE TO THESE NATURAL EXPERIMENTS

Enough was enough. Three major air pollution episodes resulted in incontrovertible evidence that exposure to extremely high air pollution concentrations, even for four or five days, could result in human (and animal) disease and death. But what could be done about it?

In the United Kingdom and the United States, attitudes about air pollution began to change. People were losing faith in the doctrine of laissez-faire, at least regarding air pollution. Air pollution no longer "smelled like money" or was an "unavoidable consequence of prosperity." It was something unhealthy that should be better understood and controlled.

Indeed, evidence of the severe health effects of air pollution from Meuse Valley, Donora, and London motivated public policy responses to protect human health and welfare. In the United Kingdom, the first Clean Air Act was introduced in 1956. In 1961, the National Survey, a coordinated national air pollution monitoring network, was established. Over time, networks that monitored particulate matter, sulfur dioxide (SO_2), nitrogen dioxide (NO_2), ozone (O_3), and other pollutants were developed and expanded.

The first federal legislative efforts dealing with air pollution in the United States began with the 1955 Air Pollution Control Act. Later the 1963 Clean Air Act and the 1967 Air Quality Act offered funds for air pollution research and launched federal efforts to monitor and control air pollution. The most crucial development in air pollution control in the United States was the Clean Air Act of 1970. A major provision of this act established National Ambient Air Quality Standards that were "requisite to protect the public health" with "an adequate margin of safety." The Clean Air Act also required regular reevaluation of these national standards based on reviews of the latest scientific evidence.[16]

The U.S. Environmental Protection Agency (EPA) was created by executive order in December 1970 with the task of implementing and maintaining the requirements of the Clean Air Act. John Bachmann, who worked with the U.S. EPA Office of Air Quality Planning and Standards, wrote an authoritative history of the air quality standards and the imperfect but beneficial standards-based approach to air quality management.[17]

There has never been political unanimity in the United States regarding the appropriate role of government in dealing with air pollution. The early 1970s, however, was a remarkable time. There was enough public support and political will to pass the Clean Air Act, establish National Ambient Air Quality Standards, create the EPA, and embark on a tumultuous journey trying to protect the air we breathe.

CONCLUSION

"Does air pollution smell like money or disease?"

Figuratively, both. Air pollution smells like money for those making profits in economic activities that pollute. Air pollution is a *cost of production* not fully paid for by the polluter. It is easier to make money if you can keep the revenues (internalize the benefits) and pass off some costs (externalize the costs).

The three major air pollution episodes discussed in this chapter presented compelling evidence that air pollution can have substantial costs to human health and well-being. To breathers in the general community, the pollution smells like costs imposed upon them. Often, these costs are hidden or not fully understood. Too often, costs include excess disease and death borne by community members breathing polluted air.

Public policy controversies, and even some scientific controversies regarding air pollution and health, are at least in part related to issues of who gets the benefits and *who pays the costs of air pollution.*

3

DOES REGULAR EXPOSURE TO AIR POLLUTION HARM POPULATION HEALTH?

As chapter 2 demonstrated, extreme air pollution events associated with temperature inversions can cause acute harm to people's health. But what about less intense, more common exposures?

Again, a natural experiment helped to answer that question. Arden took advantage of this unique natural experiment, beginning an unplanned journey in air pollution and health research.

NATURAL EXPERIMENT: UTAH VALLEY'S GENEVA STEEL

Geneva Steel was built in Utah Valley during World War II, using U.S. federal funds to increase wartime steel production. Construction began in 1941 and was completed by the end of 1944. Geneva Steel was a large, integrated steel mill with coke ovens, blast furnaces, open-hearth furnaces, rolling mills, and more. At its peak, it was the largest steel producer in the western United States. The mill was built in the Utah Valley of central Utah because of its proximity to coal, iron ore, limestone, and fresh water from Utah Lake. It was also far enough inland to avoid possible wartime attacks. The mill operated as a U.S. government facility until 1946 when it was sold to the United States Steel Corporation (U.S. Steel).

The steel mill took its name from a neighboring lakeside resort called Geneva Resort. But Geneva Steel was no resort. It was a large industrial site that produced millions of tons of steel. It also emitted immense amounts of air pollution into Utah Valley, home to residents of the Provo-Orem metropolitan area, approximately forty miles south of Salt Lake City. Air pollution from the steel mill was highly visible, and sometimes, the mill and the local area were shrouded with unsightly air pollution. Figure 3.1 is a photo of air pollution researchers standing with Geneva Steel in the background. The mill is largely obscured by its hazy pollution.

Utah Valley commonly experiences low-level temperature inversions, especially during winter months. During those inversions, local air pollution, including the pollution from the mill, would get trapped in the stagnant air near the valley floor. However, concentrations were significantly lower than those in the dramatic historical smog episodes discussed in chapter 2.

Figure 3.1
Air pollution researchers and workshop participants in front of Geneva Steel, March 19–20, 1992. Participants include (from left to right) Joel Schwartz, Bart Ostro, Arden Pope, Jane Koenig, Jack Spengler, Fred Lipfert, Sverre Vedal, Doug Dockery, and David Bates.

Geneva Steel operated continuously from 1944 until 1986. It temporarily shut down on August 1, 1986, due to a labor dispute and reopened a little over a year later, on September 1, 1987, with a change in ownership. This intermittent operation of the steel mill resulted in a unique natural experiment.

In the late 1980s, about 225,000 people lived in Utah Valley—a natural exposure chamber. The air pollution from the steel mill was shut off for thirteen months and then turned back on. Did the thirteen-month reprieve from the air pollution emitted by the steel mill impact the health of people living in the valley?

After the steel mill reopened, residents of Utah Valley talked about how much cleaner the air was while the mill was shut down. Some observed in local newspaper articles and letters to the editor that the mill's closure resulted in cleaner air and less illness. The anecdotal evidence begged for more formal and rigorous analyses.

At the time, Arden was a professor at Brigham Young University, located at the heart of Utah Valley and thus affected by Geneva Steel. After the steel mill reopened, Arden was teaching a natural resource and environmental economics class that required a term paper. One day after class, a student asked for suggestions about a research topic. Arden suggested that the student conduct a simple retrospective analysis of respiratory hospital admissions before, during, and after the mill shutdown period. Arden contacted Utah Valley Regional Medical Center, the largest local hospital in the valley, and the hospital agreed to share de-identified monthly respiratory hospital admissions data.

There were two initial problems with the timing of this study. First, in 1987—during the Geneva Steel shutdown—the EPA changed the particulate matter standards. The EPA discarded standards based on total suspended particles (TSP) and established a new standard for smaller inhalable particles (PM_{10}, particles ≤ 10 micrometers in aerodynamic diameter). This change was based on inhalation studies showing that particles larger than 10 micrometers were too large to penetrate the respiratory system past the nose and mouth and into the lungs. Population-based cross-sectional studies found stronger pollution-mortality associations with smaller particles than with larger particles (discussed in chapter 4).

Luckily, in anticipation of the new PM_{10} standards, the Utah Department of Health began monitoring PM_{10} in Utah Valley starting in the spring of 1985, providing consistent and high-quality PM_{10} data before, during, and after the steel mill's closure.

The second problem regarding timing is familiar to university professors. Between the time that Arden recommended this research project to his student and when he received the hospitalization data, the student who would use the data dropped the class.

So, out of curiosity, Arden analyzed the data himself. He had no formal training in epidemiology, public health, or medicine, but he was trained in applied statistics and econometrics. And he knew the value of a novel natural experiment. He added data from other local hospitals, conducted a careful analysis, submitted the study for peer review, and ultimately published the results in the *American Journal of Public Health*.[1]

For Arden, the Geneva study was initially a side project, an interesting natural experiment that demanded analysis simply because the research design was irresistible. Little did he know that the results and subsequent controversy would dramatically alter his research agenda and academic career.

What were the results? Children's hospital admissions for respiratory conditions were approximately twice as high during the winter months when the steel mill was operating versus the winter months when it was closed. Although Arden used statistical models to analyze the data more formally, the essential findings are easily observed by simply graphing the data.

Figure 3.2 illustrates the large and obvious reduction in particulate matter air pollution *and* children's respiratory hospitalizations when the mill was shut down. These remarkable findings classified the early Utah Valley hospitalization studies as breakthrough evidence of the health dangers of moderate air pollution. Later, during the intense debate over establishing new air quality standards for $PM_{2.5}$, the *New York Times* published an article with the headline "Utah Mill Lies at Heart of Fight for Air Pollution Limits."[2]

When the research results were published, Geneva Steel owners and executives were irritated and defensive that this research implicated the mill as contributing to illness and disease in the local community. They

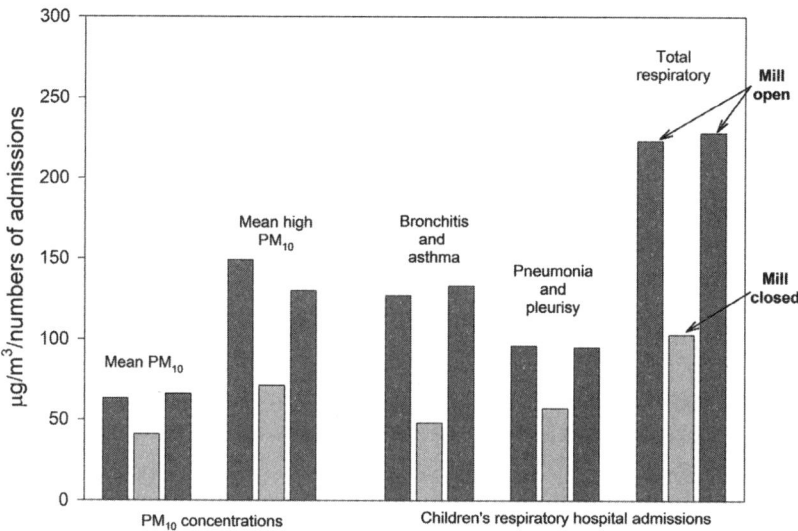

Figure 3.2
Children's respiratory hospital admissions, Utah Valley, for fall and winter months for 1985/86, 1986/87, 1987/88. *Source:* Illustration created using data from C. Arden Pope, "Respiratory Disease Associated with Community Air Pollution and a Steel Mill, Utah Valley," *American Journal of Public Health* 79, no. 5 (1989): 623–628.

were sensitive to the increasingly vocal concerns regarding the mill's air pollution. The co-owner and chair of the mill, Joseph A. Cannon, planned to run for the U.S. Senate in 1992. Geneva unleashed a barrage of reactive criticism regarding the recently published study. To debunk the research, they hired a consultant researcher, Steven Lamm, who gave local presentations arguing that Arden's published hospitalization study was flawed.[3]

Acrimonious criticism and debate were especially intense during the late summer and fall of 1989. Arden was repeatedly asked to respond to claims that his research was faulty. On September 1, 1989, Geneva Steel scheduled a press conference to publicize Lamm's criticisms of the study.

Brent Morris, a local county commissioner, and Sam Rushforth, chair of Utah Valley Citizens for Clean Air, considered the actions of Geneva Steel to be heavy-handed and deceptive. They arranged for a second or alternative press conference to be held at the county commission chambers immediately following Geneva Steel's scheduled conference.

Although the previous day, Arden had participated in an arduous forty-four-mile wilderness trail run in the High Uinta mountains, he was persuaded to present at this second press conference. Brent and Sam argued that a presentation at the press conference would be the least distracting and most effective way to respond to Lamm's assertions. Arden agreed, willing to discuss only the recently published research. Larry St. Clair, a Brigham Young University colleague and biology professor, gave clear advice: keep the presentation and response simple.

Only a few media members arrived at the second press conference's scheduled start time because Geneva Steel's press conference ran overtime. The second press conference, therefore, started late but ultimately was well attended. Arden briefly told the story of the student dropping his class, his initial analysis of the collected hospitalization data, the research methods and process, journal submission and revisions, eventual publication in the *American Journal of Public Health*, and the published research results. He noted that he was conducting further analyses of expanded hospitalization data and would eventually present the results of this extended analysis, but only after peer review and publication in a scientific journal—not at a press conference. He concluded his remarks by keeping it simple. He said that based on careful analyses of the hospitalization data, "Steve Lamm is wrong."

It isn't easy to know the effect of these dueling press conferences, but afterward, it did seem as if there was a shift in local coverage. Perceptions regarding air pollution and health research began to reflect a growing understanding that the steel mill was manufacturing more than steel. Geneva Steel's criticisms of the research were efforts to manufacture doubt and avoid being responsible for the health and environmental costs of the mill's air pollution.

Geneva Steel defended its operation by arguing that it contributed enormously to the local economy. In October 1989, they implemented a public relations campaign themed "Geneva Two-Buck Bonanza."[4] Geneva Steel employee bonuses were partially paid in two-dollar bills. The campaign was designed such that when these especially visible two-dollar bills were spent and circulated, it would demonstrate the importance of the steel mill to the local economy. In the first few days of the campaign, over $800,000 in two-dollar bills were distributed, with accompanying news coverage. While the steel mill was spotlighting its importance to the local economy, there was a continued effort to deny the environmental and public health costs of the mill's air pollution.

The contentious debate over Geneva Steel and its pollution came to an abrupt end when the steel mill became unprofitable and shut down for good in 2001. After the mill's closure, air quality in the valley improved. So did the economy. Over the next two decades, Utah Valley's Provo–Orem metropolitan area experienced remarkable population and economic growth. Utah Valley was a regional education hub, and most of the new economic growth occurred in the professional, scientific, technical, and information sectors.[5] The old site of the steel mill was cleaned up and developed for other uses. An area of Utah Valley adjacent to the former mill site has been dubbed the "Silicon Slopes" because it functions as a home to a large and growing number of successful startups and high-tech companies.[6]

Is there nostalgia for the old steel mill and its accompanying pollution? Not much. Current discussions about economic development in Utah Valley are less about trade-offs related to industrial pollution and more about the importance of education and "tech ecosystems."[7]

Even so, the Geneva Steel natural experiment contributed to the science of air pollution. In March 1992, various air pollution researchers gathered in Utah Valley to participate in a workshop addressing air pollution and health. The group posed for the photo in front of Geneva Steel (figure 3.1).

Additional analyses of the Geneva Steel natural experiment increased the evidence that the mill's air pollution substantially impacted human health. Hospitalization data from neighboring communities confirmed that excess admissions were restricted to Utah Valley.[8] Michael Ransom, a skilled Brigham Young University econometrician, collected elementary school absences in Utah Valley for six years before, during, and after the mill closure. Michael and Arden found that elevated levels of air pollution were associated with increased elementary school absences.[9]

An innovative and independent approach to evaluating the health effects of air pollution in Utah Valley was used by several resourceful EPA medical researchers and toxicologists, including Andy Ghio, Dan Costa, Bob Devlin, and Mark Frampton. They wanted to physically extract particulate matter from archived air monitoring filters from the valley for the years before, during, and after the steel mill closures (1985–1988). They then wanted to conduct controlled toxicological experiments to see if particulate matter from the filters would elicit acute airway injury and inflammation, coherent with evidence from epidemiological studies. But could they get access to the filters?

Andy Ghio was a well-trained pulmonologist who studied the effects of occupational and environmental exposures on the lungs. His medical fellowship at the University of Utah from 1986 to 1988 coincided with the Geneva Steel natural experiment. Nearly fifteen years later, Andy and colleagues made inquiries regarding archived filters from the Air Monitoring Center, Utah Division of Air Quality. Were filters available for 1985–1988 (before, during, and after the steel mill's closure)? They learned that the filters were only stored for fifteen years and then discarded. The 1985 filters were literally on the block to be trashed, with the 1986 filters to be eliminated soon after. They were just in time. Andy and colleagues arranged to have the filters from those years sent to their lab in North Carolina to undergo extraction for controlled toxicological studies.

What did they learn from these studies? They discovered that particulate matter from the filters of air pollution monitors produced acute airway injury and inflammation in rats and humans.[10] These toxicological findings were consistent with the previously reported epidemiological results.

DOUG AND ARDEN: INITIAL RESEARCH COLLABORATION

In the mid-1970s, as a graduate student in environmental health at the Harvard School of Public Health, Doug applied his engineering and physical science training to measure air pollution concentrations in the homes and personal breathing zones of people in the community. He recognized that personal exposure to air pollution was a combination of outdoor, indoor, and personal sources. Nevertheless, from a public health perspective, outdoor air pollution was most modifiable and amenable to control.

Upon graduating with a doctoral degree in 1979, Doug switched from measuring exposures to measuring health effects. He conducted several studies that measured and evaluated the impact of air pollution on lung function in children and adults (see chapter 9).[11]

In 1989, shortly after the Utah Valley hospitalization study was published, and when the controversy was still intense, Doug invited Arden to present his research from Utah Valley at the Harvard School of Public Health. The two had never met in person but were aware of each other's research. Arden was especially interested in Doug's most recently published research on the effects of particulate air pollution on children's respiratory health.[12] Doug was interested in Arden's recent Utah Valley study.[13] Their research was related, and meeting to discuss it made sense.

Arden's visit with Doug at Harvard resulted in multiple productive collaborations. Doug and his colleagues offered to help conduct studies of air pollution and respiratory health in panels of children and patients in Utah Valley. They provided peak expiratory flow meters, which monitor lung capacity. They also offered standardized questionnaires, symptom diaries, guidance, and advice.

At this time, Jack Spengler was a groundbreaking researcher in atmospheric science and environmental health and an expert in measuring environmental exposures to air pollution. He was also a co-investigator in the landmark Harvard Six-Cities Study (discussed later in chapter 4). Several researchers at Harvard, including Doug and Jack, hypothesized that acid aerosols in the air could contribute to adverse respiratory health effects. Jack offered to conduct state-of-the-art supplemental air pollution monitoring to assess Utah Valley's levels of acid aerosols.

Arden remembers the flight from Boston back to Utah. He was excited to begin using what he had learned from the Harvard researchers and started outlining research plans to conduct panel studies in Utah Valley. With Doug's collaboration and the modification of methods and materials pioneered at Harvard, the two launched a series of collaborative studies.

PANEL STUDIES AND CHILDREN'S RESPIRATORY HEALTH

The logistics and statistical methods of Arden, Doug, and Jack's collaborative panel studies were somewhat complicated, but the basic study design was simple. The Utah Department of Health conducted daily twenty-four-hour air pollution monitoring at multiple sites in Utah Valley. The research team enrolled panels of participants, mostly children, including those with and without asthma, who lived near the air pollution monitors. Participants measured and recorded their respiratory symptoms and peak flow, a measure of lung function (see figure 3.3), using standardized daily diaries and peak expiratory flow meters provided by the study. The study periods included two winter seasons. Given the multiple pollution sources (including the steel mill) and periodic temperature inversions, air pollution levels fluctuated substantially throughout the study periods.

The findings of the panel studies were remarkable. First, based on Jack's supplemental monitoring, the researchers learned that the particulate air pollution measured in Utah Valley was not acidic. This finding was a surprise because of their initial opinions that acidity was likely an important characteristic that determined the toxicity of particulate matter air pollution. Nevertheless, even without being acidic, day-to-day elevations in particulate matter air pollution were associated with reduced lung function and increased respiratory symptoms.

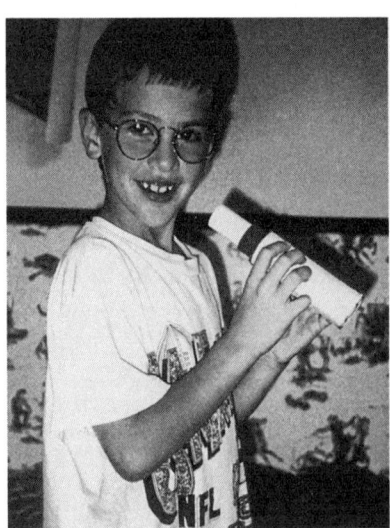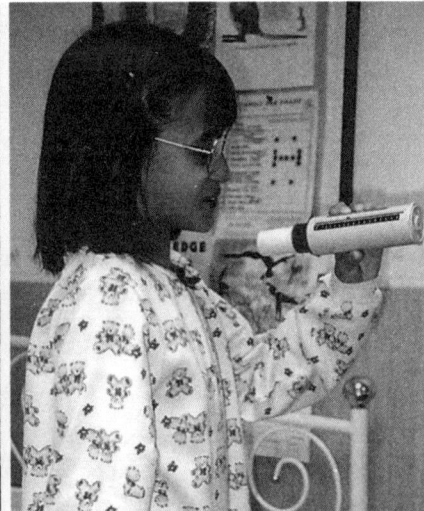

Figure 3.3
Schoolchildren measuring peak expiratory flow in 1990. *Note:* Children shown are not study participants.

Second, elevated air pollution was associated with small but measurable declines in lung function, as measured by the peak expiratory flow meters. In asthma patients, air pollution was associated with increased asthma medication use. In schoolchildren, air pollution was associated with respiratory symptoms, especially lower-respiratory symptoms and cough. Adverse effects were transient, lasting approximately five days after exposure.[14]

Studies in other areas yielded similar findings: short-term elevations in air pollution were associated with slight declines in lung function and increased respiratory symptoms.[15]

Thanks to the panel studies, scientists now had evidence suggesting that short-term exposures to low or moderate levels of air pollution contributed to respiratory illnesses. But there was emerging evidence that these low to moderate levels of air pollution contributed to much more severe health end points, including premature deaths.

EARLY MORTALITY STUDIES

Joel Schwartz received his PhD in theoretical physics. Before becoming a professor of environmental epidemiology at the Harvard School of Public Health, Joel worked as a senior scientist and econometrician for the U.S. EPA (doing landmark analysis of the health and economic benefits of reducing lead in gasoline). A highly productive air pollution researcher, he applied innovative strategies to analyze health and pollution. He analyzed the number of deaths per day compared with air pollution data from London for fourteen winters, an early example of what would be called daily time-series mortality studies. He demonstrated that mortality in London was associated with air pollution—not just during extreme episodes but even at low to moderate pollution levels.[16]

Bart Ostro, trained as an economist, was an important and steady contributor to the literature on the health effects of air pollution for over forty years. He also analyzed the expanded London data with findings similar to Joel Schwartz's work.[17]

The EPA, however, discounted studies using London data in setting air quality standards, saying they did not apply to U.S. populations. Similar time-series studies were not feasible in most U.S. cities because particulate air pollution data were collected every sixth day, making analyses of daily associations impossible.

Fortunately, daily particulate air pollution measures were available for some cities, including the relatively small city of Steubenville, Ohio, and the larger city of Philadelphia, Pennsylvania. In 1991, Joel Schwartz was on sabbatical at Harvard. In collaboration with Doug, Joel applied innovative time-series methods and demonstrated that daily deaths were associated with daily particulate air pollution levels in both Steubenville and Philadelphia.[18]

While conducting these initial analyses of daily mortality in Steubenville and Philadelphia, Joel called Arden. Joel noted that a recent study by David Fairley observed that daily changes in particulate air pollution were associated with daily mortality in Santa Clara County, California.[19] Joel said he and Doug detected similar associations with particulate air pollution and daily mortality counts in Steubenville and Philadelphia. Joel wondered, What about Utah Valley? He knew that daily air pollution was monitored there. When he asked Arden if they could access local mortality data and relate air pollution to deaths for various causes in Utah Valley's metro area, the answer was yes. Such data were available from the Utah State Department of Health's vital records.

Later, Joel provided Arden with an individual tutoring session. Joel and his wife, Ronnie Levin, an accomplished EPA researcher, took a snow-skiing vacation to a resort in Utah. One evening, Arden drove up to the ski lodge where they were staying. He and Joel discussed statistical approaches to analyze the daily mortality data. Joel used a piece of scratch paper from the ski lodge and outlined the basic code needed for the statistical analysis.

Within two years, Joel had led or collaborated with Doug, Arden, and others and published formal daily time-series studies from Philadelphia; Steubenville; Utah Valley; St. Louis, Missouri; eastern Tennessee; and Birmingham, Alabama.[20] All the studies had similar results: daily changes in particulate matter air pollution were associated with changes in daily mortality counts.

These daily mortality time-series studies had two essential and troubling findings: (1) the pollution-mortality associations extended to cardiovascular deaths, and (2) the pollution-mortality associations occurred at lower air pollution levels than what was set by the National Ambient Air Quality Standards. The results suggested that air pollution had a more extensive health impact than expected and that current air quality standards were inadequate to protect public health.

There was controversy and criticism regarding these studies. Two lively conferences were conducted in Irvine, California, in January 1994[21] and in Park City, Utah, in May 1996.[22] A diverse group of researchers and regulators interested in particulate air pollution epidemiology came together. Industry-supported investigators attended and presented analyses directly challenging the results of the early time-series analyses.[23] Some asserted that they could not reproduce the data or the reported results. Some suggested that the associations might be artifacts of the statistical methods used and thus would be misleading. Some implied that weather variables were not adequately addressed. Some suggested that the daily pollution-mortality association was only mortality *displacement*. In other words, they claimed that air pollution only moved the date of death ahead by a few days. This phenomenon was sometimes less sensitively referred to as "harvesting" of frail people who were about to die anyway.

The debate surrounding the validity of these studies was intensified by a lawsuit against the U.S. EPA by the American Lung Association, calling for a review of the National Ambient Air Quality Standards. The lawsuit resulted in a court order that sped up the assessment of the science that supported the particulate matter limits in the National Ambient Air Quality Standards.

GOING DEEPER: REPLICATION AND ADVANCED STUDIES

Dissent in science may sound unpleasant, but it is entirely normal. Indeed, the scientific process depends on researchers questioning others' findings and attempting to replicate them independently. Efforts to address criticisms and controversies often improve the science in a field and help refine methods.

For example, Larry Kalkstein, a well-respected climatologist, was critical of the early daily time-series mortality studies. He argued that the initial statistical modeling approaches used to adjust for weather variables—temperature, wind, precipitation—were inadequate. These methods allowed for flexible season, temperature, and relative humidity modeling. Larry argued that the researchers should use a "synoptic weather modeling" approach that used more meteorological variables and allowed for control of more complex weather conditions. He suspected this approach would eliminate or reduce the estimated adjusted daily associations between air pollution and mortality.

In January 1995, Doug, Arden, Joel Schwartz, Larry Kalkstein, and others made presentations at meetings sponsored by the EPA at Research Triangle Park, North Carolina. While at the presenter's table, as part of a formal panel discussion, Larry and Joel got into a heated debate about which of the two modeling approaches worked better. Arden happened to be sitting between them, mostly keeping his head low. When there was a brief break in the action, Arden asserted that the debate over the best weather model was an empirical question. He had adequate data from the Wasatch Front area of Utah to try both modeling approaches. He bet Larry a milkshake that the method used by Joel, Doug, and Arden would work better than Larry's approach.

At the end of the session, as they left the meeting room, Larry and Arden were intercepted by EPA personnel. They asked if they would work together to compare the two approaches using a common data set. Allan Marcus, an experienced EPA statistician and epidemiologist, was especially encouraging and supportive of this collaborative effort. Although their prior interactions had been mostly adversarial, Larry and Arden agreed to work together. The EPA provided essential funding to conduct the research. Arden flew out to the Center for Climatic Research at the University of Delaware and worked with Larry to compare the two modeling approaches.

Who won the milkshake? It was a toss-up. Both models fit the data equally well. Further, Larry's alternative additional weather variables did not affect the estimated pollution-mortality associations.[24] Larry and Arden had a short but productive collaboration. In the end, no one bought milkshakes—but they had a nice dinner together. Consistent with the finding that their models worked equally well, they split the check. In the aftermath, Larry collaborated with another research team using different data sets but with similar results.[25]

A much more crucial effort to address controversy and criticisms occurred when the Health Effects Institute (HEI) sponsored a reanalysis of the early daily time-series mortality studies. The HEI was chartered in 1980 as an independent research organization funded by both the EPA and industry, primarily the worldwide motor vehicle industry. The intent of HEI is to provide impartial, high-quality, and relevant science on the health effects of air pollution. An expert oversight committee, appointed by the HEI board of directors, selected a team of scientists to conduct the reanalysis and further expand the analysis. The independent research team was led by two notable researchers: Jonathan Samet, a pulmonary physician and highly respected environmental epidemiologist who had trained with Frank Speizer and Ben Ferris at Harvard; and Scott Zeger, a skilled, innovative statistician. Both were faculty members at Johns Hopkins School of Public Health. Several investigators of the early studies, including Joel, Doug, Arden, and David Fairley, cooperated with the reanalysis project and provided their data files. In August 1995, the independent researchers reported replicating and validating the findings reported by the original investigators.[26]

The HEI replication and validation of these early single-city time-series studies marked a turning point for the time-series studies. Throughout the next ten years, over a hundred time-series studies were published in the peer-reviewed literature.

Additionally, extensive, coordinated multi-city time-series studies provided crucial scientific contributions. One was the HEI-sponsored National Morbidity, Mortality, and Air Pollution Study (NMMAPS). Francesca Dominici, a dynamic and creative "big data" statistician, joined Jon Samet and Scott Zeger at Johns Hopkins. They conducted analyses with combined data for up to 100 cities in the United States, documenting day-to-day changes in particulate matter air pollution associated with day-to-day changes in daily death counts.[27]

Klea Katsouyanni, a medical statistics and epidemiology professor at the University of Athens Medical School, coordinated and led another large multi-city study. This study, Air Pollution and Health: A European Approach (APHEA), combined data from twenty-nine European cities.[28] Klea and Jon Samet then joined forces with their research teams to analyze combined data from cities in Europe and North America. This study was called Air Pollution and Health: A Combined European and North American Approach (APHENA).[29] These ambitious multi-city studies provided statistically powerful pooled measures of association across many cities. They provided opportunities to evaluate exposures from broad geographic areas with diverse exposures and sources. They further developed the analytic underpinnings of the daily time-series approaches. These multi-city findings undermined previous assertions that London air pollution studies were not relevant to U.S. cities.

One of the most remarkable multi-city studies of daily air pollution and mortality looked at 652 cities worldwide with adequate pollution and mortality data.[30] This study plotted pooled mortality changes against changes in $PM_{2.5}$ for all 652 cities. As illustrated in figure 3.4, mortality risk was elevated when air pollution concentrations rose. When this study was published, John Balmes, professor of medicine at the University

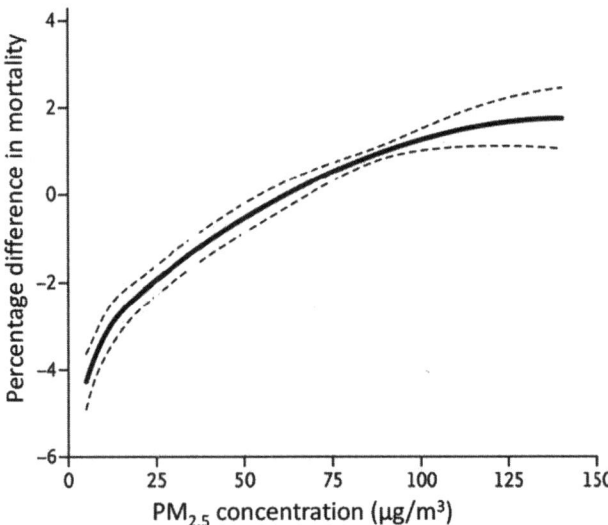

Figure 3.4
Ambient particulate air pollution and daily mortality in 652 cities. Pooled concentration-response function. *Source:* Adapted from C. Liu, R. Chen, F. Sera, A. M. Vicedo-Cabrera, Y. Guo, S. Tong, M. S. Z. S. Coelho, et al., "Ambient Air Pollution and Mortality in 652 Cities," *New England Journal of Medicine* 381, no. 21 (2019): 2072–2075. Reprinted with permission from Massachusetts Medical Society.

of California at San Francisco and highly respected expert on the health effects of air pollution, wrote an editorial titled: "Do We Really Need Another Time-Series Study of the $PM_{2.5}$-Mortality Association?"[31] He had a point.

There was now a great deal of evidence linking air pollution with death. In addition to mortality, time-series studies were applied to non-fatal health events such as hospital admissions.[32] The Johns Hopkins research team, including Francesca Dominici, Jon Samet, and Scott Zeger, conducted extensive daily time-series studies of hospitalizations. They analyzed data on hospitalizations in 204 U.S. urban counties, using data from the Medicare Nation Claims History Files. Their conclusion was straightforward: "Short-term exposure to fine particulate matter air pollution ($PM_{2.5}$) increases the risk for hospital admission for cardiovascular and respiratory diseases."[33]

INNOVATIVE METHODS: CASE-CROSSOVER STUDIES

Another approach to help confirm evidence from daily time-series studies is to utilize new analytic strategies from other fields. For example, Malcolm Maclure, an innovative epidemiologist in research methods at the Harvard School of Public Health, developed a *case-crossover design* to study the transient health effects of various acute events.[34] This design allowed researchers to study the effects of short-term exposures to intense air pollution on the risk of a well-defined adverse health event (such as death, heart attack, or acute respiratory distress). Exposures at the time of, or shortly before, the event are matched with exposures when the event did not occur (these would be the control periods). The excess risk is evaluated using a rigorous statistical approach called *conditional logistic regression*.

The case-crossover design offers some advantages. Study subjects serve as their own controls (matched with themselves at other times); cross-subject differences that change slowly over time (such as smoking history, body mass index, diet) are controlled by design, with less reliance on statistical modeling. Also, if matching control periods are close in time and on the same day of the week, then day-of-week, seasonality, and long-term time trends are controlled by design rather than by statistical modeling. As such, this design mimics a controlled experiment and is "quasi-experimental."

Teaming with colleague Murray Mittleman, a Harvard cardiologist and epidemiologist, Maclure applied the case-crossover approach to studies of the risk of acute myocardial infarction (heart attack). They evaluated heart attack risk with exposure to activities including heavy physical exertion, episodes of anger, sexual activity, and use of cocaine and marijuana.[35]

The case-crossover design also was used to study short-term air pollution exposures with the risk of mortality using data from Philadelphia; Seoul, Korea; and a study of fourteen U.S. cities.[36] The findings were consistent with the daily time-series studies.

The case-crossover analytic approach was further used to evaluate how short-term exposure to air pollution affects nonfatal events. Annette Peters, an environmental epidemiologist at Helmholtz Zentrum Munchen in Germany, previously studied with Doug at Harvard as a visiting doctoral student. After she received her degree, she returned for a year and initiated a collaboration with Murray Mittleman and Doug to link heart attacks with hourly or daily measures of $PM_{2.5}$. Using the case-crossover approach, they found that the risk of heart attacks was elevated a few hours to one day after exposure to $PM_{2.5}$ air pollution.[37]

However, the case-crossover design still had one issue: What was the best strategy for selecting the control periods? Lianne Sheppard, a versatile biostatistician at the University of Washington, and her colleagues proposed a time-stratified control selection approach that matched control periods on days and months of the events. For example, if the event occurred on a Tuesday in the month of May, the control periods would include all the other three or four Tuesdays of that same month. They demonstrated the efficacy of this relatively simple and straightforward approach, and it became a standard method.[38]

In a massive case-crossover analysis of short-term exposure to air pollution and mortality, researchers from Harvard drew data from the entire U.S. Medicare population from 2000 through 2012. The results? Air pollution, including $PM_{2.5}$ and warm-season ozone, was associated with mortality risk. The pollution-mortality association persisted *even below* U.S. National Ambient Air Quality Standards levels.[39]

Arden was asked to write an editorial about using the case-crossover design in the early mortality and air pollution studies.[40] He was familiar with Doug and Annette's study and appreciated the quasi-experimental aspect of the design. One day, he read a paper on white blood cell subtypes and cardiovascular risk.[41] The study had nothing to do with air pollution but used participants drawn from a cardiac catheterization registry of the Intermountain Heart Collaborative Study. This study offered data on the timing of acute ischemic heart disease events (acute myocardial infarction and chest pain known as angina). It had excellent data on the research subjects, including age, gender, and smoking status. The subjects also underwent coronary angiography, providing information on the number of diseased coronary vessels.

When Arden read this study, he realized that data from these registry patients could be used in a novel case-crossover study of short-term air pollution.

The cardiologists who conducted the original research were at the Intermountain Medical Center cardiovascular department, just fifty miles up the road near Salt Lake City. They agreed to meet with Arden and ultimately began a collaboration with him.

The team linked data for nearly 13,000 research subjects who lived in Utah's populated Wasatch Front with monitored air pollution data. They analyzed the data using a case-crossover design. They found that short-term exposures to $PM_{2.5}$ were associated with an increased risk of acute coronary events (heart attacks and unstable angina), especially in those with angiographic evidence of existing coronary artery disease.[42]

Additional collaborative studies by Arden and the research team at Intermountain Medical Center found that short-term exposure to $PM_{2.5}$ air pollution also increased the risk of heart failure hospitalization, acute heart disease events, and healthcare visits for acute lower respiratory infections.[43]

Other research teams have used the case-crossover approach to study very short-term exposures. For example, a team of researchers, consisting primarily of cardiologists and pulmonologists at the University of Rochester Medical Center, found that elevated concentrations of $PM_{2.5}$ air pollution one hour before the event were associated with increased risks of specific types of heart attacks.[44] A systematic review and meta-analysis of air pollution and heart attacks[45] demonstrated that particulate air pollution is significantly associated with the risk of heart attacks in both case-crossover and daily time-series studies. Recently, an extremely large Chinese study found that even hourly elevations in air pollution were associated with the onset of acute coronary syndrome, including heart attacks.[46]

CONCLUSION

"Does air pollution exposure at common levels have adverse health effects?"

Yes. The evidence indicates that exposure to elevated air pollution is associated with the following:

1. Small transient declines in lung function
2. Increases in respiratory symptoms, especially lower-respiratory symptoms, cough, and trouble breathing
3. Increases in school absences
4. Increased risk of acute cardiovascular disease events, including heart attacks
5. Increased risk of acute respiratory events
6. Increased risk of hospitalizations for both respiratory and cardiovascular disease
7. Increased risk of death for both respiratory and cardiovascular disease.

All this evidence of the adverse health effects of air pollution comes from studies that evaluated only relatively short-term changes in pollution exposure—only days or months. But what are the health effects of longer-term exposure to air pollution—for years or decades?

4

DO LONG-TERM EXPOSURES INCREASE
THE RISK OF DISEASE AND DEATH?

Most of the studies outlined in the previous chapters looked at short-term exposure, whether to intense air pollution or everyday pollution levels. But people who live or work in polluted areas absorb air pollution over their lifetime. What are the health costs to those people?

DEATH RATES FROM LONG-TERM EXPOSURES ACROSS CITIES

As an environmental economist at Carnegie Mellon University, Lester Lave became one of the university's most distinguished and well-respected professors. In 1970, early in his career, Lave, with one of his graduate students, Eugene Seskin, published an audacious paper in the journal *Science* titled "Air Pollution and Human Health."[1] Lave and Seskin presented analyses using data from 114 metropolitan statistical areas in the United States. Using statistical models that adjusted for available socioeconomic variables, they showed that mortality rates were significantly associated with particulate air pollution (including sulfate particles). Their results suggested that air pollution is likely an important contributor to human disease and death. Lave and Seskin later published a book presenting approximately ten years of extended analysis that evaluated the association between mortality rates and air pollution.[2]

Lave and Seskin's initial *population-based cross-sectional studies* prompted substantial scrutiny and spawned similar analyses by other researchers using multiple data sets and periods.[3] A notable study demonstrated that mortality rates were most strongly associated with sulfate and fine particulate matter.[4] Prominent researchers, including John Evans, a Harvard environmental science professor and expert in risk assessment, argued that these studies were offering meaningful insights into the real health impacts of air pollution, rather than merely spurious correlations, and should be taken seriously.[5]

However, these studies on mortality rates and air pollution were also highly criticized. Some cast doubt on their usefulness to public health policy decisions. Their argument was that the observed correlations between air pollution and mortality rates were not causal but were just the result of other characteristics of these metropolitan areas, regardless of statistical modeling efforts to control for them. Some noted that population-based data could create an ecological or population fallacy. They asserted that the only way to be confident that long-term exposure contributes to human disease and death requires long-term longitudinal studies based on data from large cohorts of individuals. Ideally, researchers should use "quality prospective cohort studies," where individuals living in areas with different pollution levels are followed for long periods and then analyzed.

The problem with quality prospective cohort studies is that they are big, expensive, challenging, and take many years to complete. Who would conduct such a study? In the early 1970s, the EPA attempted similar research, called the Community Health and Environmental Surveillance System (CHESS) studies. The studies were challenged based on concerns regarding design, data analysis, and even reporting.[6] For a time, the EPA refrained from conducting additional epidemiological research. Following the 1973 oil embargo, anticipating significant changes in levels and sources of air pollution, the National Institute of Environmental Health Sciences (NIEHS) invited research proposals to study the health effects of air pollution.

TAKING THE LONG VIEW: THE HARVARD SIX-CITIES STUDY

Benjamin Ferris, a pediatrician by training, was drawn to teaching and research on respiratory disease. He was interested in the risk factors that contribute to this disease. In the early 1970s, Ferris was a Harvard professor of environmental health and a leading expert in field epidemiology. He was also an avid and accomplished mountaineer. His colleague, Frank Speizer, a Harvard professor of medicine and environmental science, was also an energetic, enthusiastic, and respected researcher. Speizer was dedicated to understanding the natural history of respiratory disease and environmental risk factors.

Ferris and Speizer responded to the NIEHS's request for research proposals. They proposed an ambitious *prospective cohort study* that is now commonly referred to as the Harvard Six-Cities study. Their plan was to study respiratory health effects of particulate air pollution and sulfur oxides on adults and children living in six U.S. cities (Portage, Wisconsin; Topeka, Kansas; Watertown, Massachusetts; Kingston, Tennessee; St. Louis, Missouri, and Steubenville, Ohio). The study was funded, and enrollment of research participants began in 1974.

Benjamin Ferris and Frank Speizer initiated and led the Harvard Six-Cities study, but with a research project of this magnitude, they needed help. Jack Spengler, a young Harvard environmental science and engineering professor, joined the team. He designed monitoring systems and methods and conducted air pollution monitoring. Doug joined the team as a graduate student and later, in 1988, became the principal investigator of the study.

Although there were multiple components to the study, a core element was enrolling samples of adults from six cities with a wide range of air pollutants. Participant information regarding age, sex, weight, height, education level, smoking history, occupation exposures, diet, and other relevant factors was collected at enrollment.

Each year the team checked for changes of address or death. The study's objective was to evaluate the potential impact of air pollution. The plan was to conduct mortality follow-up for a sufficient period and then conduct statistical survival analyses using the collected data.

In 1992, Arden was a visiting scientist and an Interdisciplinary Program in Health Fellow working with Doug and colleagues at the Harvard School of Public Health. Doug and Frank Speizer made Arden a generous offer. At that time, they had collected up to sixteen years of mortality follow-up with relevant air pollution and other data for the Harvard

Six-Cities study. They asked if he would help analyze and interpret the data. Although Arden had not helped collect the data, he jumped at the opportunity to explore data from such a unique and remarkable study.

Analysis of the Harvard Six-Cities mortality study was facilitated by its elegant study design and its high-quality, carefully documented, and well-formatted data. The results were fascinating. The team calculated survival probabilities for each year of follow-up and plotted survival probability curves for each of the six cities. The chances of survival in each were significantly different, with the participants in the more polluted cities dying more rapidly.

The team used statistical survival models to estimate the relative risk of mortality associated with various air pollutants while adjusting for age, sex, smoking history, education levels, body mass index, and occupational exposures. Death risk was not strongly associated with aerosol acidity or ozone, but it *was* significantly related to particulate matter and sulfur air pollution—especially $PM_{2.5}$. The association between mortality risk and $PM_{2.5}$ is illustrated in figure 4.1. $PM_{2.5}$ mortality associations were observed for all-cause mortality and combined respiratory and cardiovascular disease deaths but not for other causes of death.

As the team began to get results from the data analysis, they became concerned. The effects of air pollution on mortality risk were much larger than expected based on the daily time-series mortality studies. The results showed that those living in Steubenville, the most polluted city, were dying at a rate 26 percent higher than those in the least polluted city, Portage. The pollution-mortality association was strongest with fine particles and was nearly linear (see figure 4.1). The team discussed these unexpectedly strong associations between air pollution and mortality risks. They tried different statistical models and evaluated the sensitivity of the results. They analyzed the data using various stratifications. They tried excluding subjects with hypertension or diabetes. The results were remarkably robust; that is, the basic results were highly consistent regardless of the various ways of analyzing the data. Nothing the researchers did substantially changed the findings.

The results indicated that air pollution in the United States had substantial adverse effects on life expectancy, even in cities that met the current ambient air quality standards. The team knew that the results were

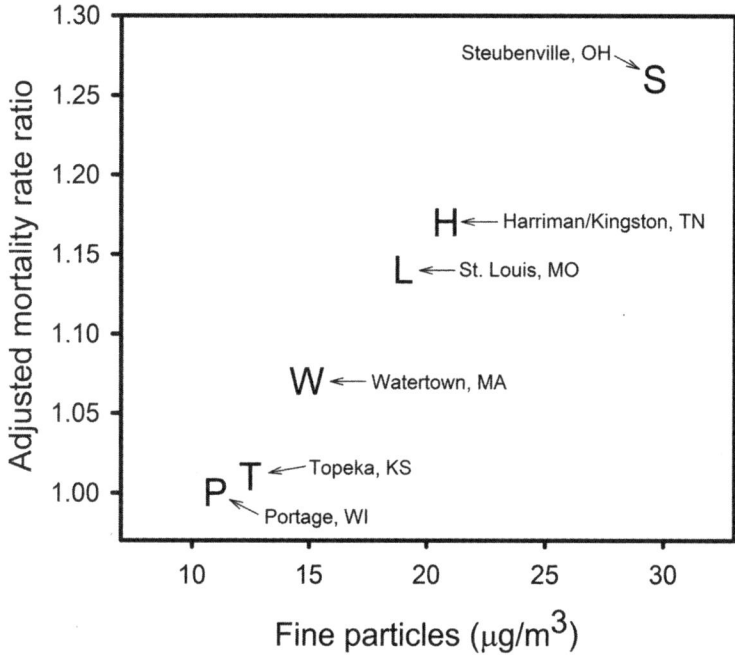

Figure 4.1
Estimated adjusted mortality rate ratios plotted over mean concentrations of PM$_{2.5}$ from the original Harvard Six-Cities study. *Source:* Adapted and replotted from results reported in Douglas W. Dockery, C. Arden Pope, Xiping Xu, John D. Spengler, James H. Ware, Martha E. Fay, Benjamin G. Ferris, and Frank E. Speizer, "An Association between Air Pollution and Mortality in Six U.S. Cities," *New England Journal of Medicine* 329, no. 24 (1993): 1753–1759.

important and should be reported. However, they were cautious because the estimated pollution-mortality association was unexpectedly strong. Could the results be replicated in an alternative, independently collected cohort? And how could one replicate a prospective cohort study that required nearly twenty years to plan, fund, initiate, and conduct? Frank Speizer suggested that they contact the American Cancer Society.

FOLLOWING UP: THE AMERICAN CANCER SOCIETY COHORT

In 1982, the American Cancer Society (ACS) began an ambitious prospective cohort study called the ACS Cancer Prevention Study II (CPS-II). They enrolled 1.2 million adults throughout the United States and followed them over time to learn at what age they died and what caused their death. Upon entering the study, participants provided information about age, sex, race, height, weight, smoking history, alcohol use, diet, and more. The ACS used this massive cohort to study cancer risk factors. They followed up regularly to track participants' dates and causes of death. The Harvard Six-Cities study team, including Doug and Arden, realized they might use the ACS CPS-II cohort data to replicate and reevaluate their air pollution results.

Frank Speizer initiated a collaboration with ACS researchers. The initial ACS collaborators included Clark W. Heath Jr., head of the ACS department of epidemiology and statistics, and Michael J. Thun, who succeeded Clark Heath in 1998. Subsequent ACS collaborators included Eugenia E. Calle, Susan Gapstur, and Ryan Diver. These ACS collaborators were incredibly generous. They facilitated collaborative analyses of their data while carefully protecting the confidentiality of the cohort participants. They contributed remarkable skills, expertise, and an unflinching willingness to pursue the evidence. Collaborations with the ACS continued for much longer than expected—and through circumstances that were sometimes difficult and controversial.

A key difference between the Harvard Six-Cities study and the ACS CPS-II cohort was that the ACS had not designed its cohort as an air pollution study. Therefore, a primary challenge of using ACS CPS-II cohort data was assigning air pollution exposure estimates to cohort participants. Based on the results from the Harvard Six-Cities study, the research question for the ACS study was clear: Was $PM_{2.5}$ and sulfate air pollution associated with mortality risk, especially cardiopulmonary and lung cancer mortality?

The combustion of fossil fuels generates both fine ($PM_{2.5}$) and sulfate particles. Because sulfate particles are a sizable fraction of fine particulate matter, both pollution measures were highly correlated and represented combustion-source air pollution. Metro-level air pollution estimates for $PM_{2.5}$ were available for fifty metropolitan areas. Estimates for sulfate particles were available for 151 metro areas. The team linked these estimates with the CPS-II cohort at the metro level.

The team analyzed the data using statistical survival models, estimating the risk of death associated with each of the two air pollutants. The models controlled for key risk factors, including age, sex, race, smoking history, alcohol consumption, exposure to passive smoke, education levels, body mass index, and occupational exposures.

This was before wide internet use and powerful personal computers, and the data analysis offered some logistical challenges. The cohort size and computational complexity of the models required travel between Boston and Atlanta, overnight computing on the secure mainframe computer in Atlanta, and careful collaboration with ACS programmers and data managers. It took extra work to link participants' addresses with air pollution records from those locations over time.

Nevertheless, the results were reasonably clear: both fine and sulfate particulate matter were associated with an increased mortality risk, especially cardiopulmonary mortality (cardiovascular and respiratory deaths combined). The results were less conclusive for lung cancer.

The ACS CPS-II study was a good match, even complementary to the Harvard Six-Cities study. Both were well-designed cohort studies, but they had different strengths and limitations. The Harvard Six-Cities study offered an elegant and balanced study design and study-specific air pollution monitoring. Further, the study's results could be presented in an easy-to-understand graphical format (figure 4.1). The Harvard Six-Cities study was limited by including only six cities and having a relatively small cohort.

The ACS CPS-II was limited by the availability of historical air pollution data from monitoring networks. However, a major strength of the ACS CPS-II cohort study was its size, with hundreds of thousands of participants distributed across many U.S. cities. The collaboration also demonstrated that epidemiological studies of air pollution could utilize other cohorts not designed to focus on air pollution. Using different cohorts opened the door for dozens of subsequent analyses (see chapter 5). Finally—and most importantly—the ACS CPS-II collaborative analysis confirmed that the pollution-mortality associations in the Harvard Six-Cities study were likely credible and not spurious.

Upon receiving the confirmatory evidence from the ACS CPS-II analyses, the team felt ready to submit the report from the Harvard Six-Cities study for publication.[7] Soon after, they submitted a manuscript reporting the air pollution results from the ACS CPS-II cohort.[8] These two primary studies quickly entered a public policy debate that was both receptive and hostile.

SCRUTINY, CONTROVERSY, AND NEW FINE PARTICULATE MATTER (PM$_{2.5}$) STANDARDS

Almost as soon as the results of the Harvard Six-Cities and ACS CPS-II air pollution studies were published, there was scrutiny, controversy, and debate. Many shared the same concerns that the collaborative team first had. These studies estimated effects of air pollution that were surprisingly large—much larger than observed from the daily time-series mortality and case-crossover studies. Critics also questioned the adequacy of the air pollution exposure data used in the analyses. Some asked whether all of the factors that could possibly be responsible for the increase in mortality (potential confounders) had been adequately adjusted. For example, in the original Six-Cities analysis, was age adjustment using five-year age groups sufficiently refined? And did those large age groups artificially influence the pollution-mortality effect estimates? In fact, when the team reanalyzed the data using one-year age groups, the estimated adjusted mortality rate ratio was unchanged.[9]

The controversy over the reported results of these two cohort studies went beyond quibbling about statistical analysis, control of confounders, or controlling for age. The results were reported during the mid-1990s, when many new studies showing the effects of short-term air pollution exposure, including the daily time-series and case-crossover mortality studies, were still being assimilated, evaluated, and replicated. The team's new prospective cohort study findings, indicating that long-term exposure to air pollution had much more significant effects on mortality risk than previously believed, changed the dynamics of the debate.

Further stoking the controversy, the two studies were published during intense and concerted efforts to review the national ambient air pollution standards for particulate matter air pollution and establish a new PM$_{2.5}$ standard. Around that time, lawsuits by the American Lung Association

(1993 and 1994) challenged the EPA to review the national ambient air quality standards faster.[10]

In 1997, the EPA proposed new national ambient air quality standards for $PM_{2.5}$ pollution and more stringent standards for ozone. The establishment of the new $PM_{2.5}$ standard and the more focused attention on fine particulate matter air pollution were the most crucial developments in the U.S. efforts to control air pollution and protect public health since the 1970 Clean Air Act. But there were some who firmly resisted these efforts.

The American Trucking Association, various private companies, and the states of Michigan, Ohio, and West Virginia filed lawsuits challenging the EPA's authority to issue these standards. These legal challenges ended up at the U.S. Supreme Court. In 2001, the court unanimously ruled that the 1970 Clean Air Act allowed the EPA to establish national ambient air quality standards to protect public health.[11] This ruling ultimately allowed the EPA to implement and enforce the new $PM_{2.5}$ standards.

The publication of these new air pollution and health studies, the debate on the quality of these studies by highly motivated and well-funded interest groups, the proposed and contested new air quality standards for $PM_{2.5}$, and the legal challenges all resulted in a political, legal, and scientific environment of consternation, debate, and sometimes outright acrimony. Jocelyn Kaiser reported in an article in *Science* titled "Showdown over Clean Air Science" that "industry and environmental researchers are squaring off over studies linking air pollution and illness in what some are calling the biggest environmental fight of the decade."[12]

The Harvard Six-Cities and ACS CPS-II cohort collaborative studies were a critical part of this acrimonious debate and were often targeted by critics.

INDEPENDENT REANALYSIS OF THE LONG-TERM STUDIES

In 1997, representatives of industry, members of the U.S. Congress, health researchers, and the EPA urged the researchers to share their data to help settle some of the controversies through independent reanalysis. In other words, allow others to analyze the same data sets. The teams involved with the Harvard Six-Cities and ACS CPS-II cohort air pollution studies agreed to make the original data available. So did Harvard University and the American Cancer Society as data owners. Data sharing was conducted under the sponsorship and oversight of the Health Effects Institute.

The HEI provided guiding principles for the reanalysis of the Harvard Six-Cities and ACS CPS-II cohort studies. For example, the reanalysis would be of the highest scientific quality, as open and public as possible, conducted by independent and impartial investigators selected by a competitive process, and subject to independent and rigorous peer review. Furthermore, the HEI would distribute a comprehensive report of all analyses and findings. HEI assembled an expert panel to provide scientific oversight of the project, and they organized an advisory board of experts from industry, academia, government, and nongovernmental organizations.[13]

The primary original investigators included Doug, Arden, Frank Speizer, and Michael Thun. Although the original investigators played no part in the reanalysis, they agreed to cooperate and provide files, computer code, and background information as needed. After the reanalysis was complete, the original investigators could respond and provide comments to be added to the final report.

To select the reanalysis team, the HEI issued "A Request for Qualifications: Epidemiologists and Biostatisticians to Design and Conduct a Reanalysis." Thirteen U.S., Canadian, and European teams responded. The HEI expert panel ultimately recommended a team of scientists, mainly from leading Canadian universities. The HEI board of directors approved this recommendation in November 1997.

Heading the reanalysis research team was Daniel Krewski, a statistician, risk scientist, and professor of epidemiology at the University of Ottawa. In addition to being an excellent researcher, he was well-respected, calm, and broad-minded. Also on the team was Richard Burnett, a mathematical and applied statistician with experience studying health risks associated with air pollution. As an indefatigable and proficient researcher, Rick became indispensable to the reanalysis project. The entire research team included thirty-three investigators from eleven North American research institutions.

The reanalysis project was conducted under a well-specified "Memorandum of Understanding" that confirmed the rights of data ownership and conditions for data access. Crucially, it preserved the confidentiality of research subjects and the integrity of the original data. HEI, the expert panel, the original investigators, and the reanalysis team signed this memorandum in March 1998. Thus began an intense two-year effort consisting of a quality assurance audit of the data, replication, and validation of the originally reported results, as well as sensitivity analyses to evaluate the originally reported results.

Doug, Arden, and the other original investigators had confidence in their work but found the reanalysis period stressful. Looking back at their original work, the team knew they had made every effort to do an excellent job conducting, analyzing, and reporting the Harvard Six-Cities and ACS CPS-II studies. They had worked with honesty and integrity. They had even hesitated over the unexpectedly large pollution-mortality effects in the Harvard Six-Cities analysis until the ACS CPS-II results confirmed them. Both studies were pioneering research efforts without clearly established standard approaches to analyze the data. There could be differences in scientific judgment about how best to analyze the data, and the results could be sensitive to some of those analytic decisions. Of course, the team hoped that the independent reanalysis would fully validate their work.

The results of the reanalysis were reported on July 26, 2000. HEI convened a symposium at the American Association for the Advancement of Science in Washington, DC. At this symposium, presentations summarizing the findings were given, and an extensive 300-page reanalysis report was released and published online.[14] At the time, the new EPA standards for $PM_{2.5}$ were embroiled in litigation. Adding to the suspense, even the original investigators did not receive the report or preview of the results until the public release.

So, what were the findings of this large, multi-year reanalysis effort? The headline of the press release succinctly summarized the findings: "New Analysis Confirms Results of Key Studies of Particles and Mortality."[15] The reanalysis team reported some minor data and analytic issues in the data audit. Still, they confirmed the data's quality and established that the initial results could be replicated and validated. They also found that the pollution-mortality associations were remarkably robust[16] and closely replicated the originally reported results.

The original investigators were relieved and thrilled. But they still had some concerns about the reanalysis. They considered both initial studies (Six-Cities and ACS CPS-II) straightforward, clean, elegant approaches to generating and testing specific, well-defined hypotheses. They responded to the reanalysis team:

Much of the elegance has been lost in the reanalysis, which at times seemed not to be hypothesis-driven at all but to be an attempt to bludgeon the data until they succumbed. In fairness, this was done very systematically and skillfully.[17]

Was the original research team too sensitive? Probably. While scientists embrace peer review and questioning of research, they are also human, and humans often dislike being questioned. The reanalysis was expansive and unrelenting. We found the level of nit-picking in the reanalysis annoying, but we respected that it was the reanalysis team's job to scrutinize their work thoroughly. The 300-page report was methodical and comprehensive. It was honest and skillfully done. It made interesting and important methodological contributions, representing a massive amount of work by a highly skilled and dedicated research team. Even if some aspects were annoying, we appreciated that the reanalysis ultimately validated the importance and scientific merit of the original studies.

EXTENDED ANALYSES OF THE ACS CPS-II COHORT

The independent reanalysis of the Harvard Six-Cities and ACS cohort studies did not end the use of these two original cohorts. However, following the release of the HEI reanalysis report in 2000 and the Supreme Court ruling in 2001, there was a reprieve in the vociferous controversies and rancor focused on the Harvard Six-Cities and ACS studies. The calm lasted for about a dozen years until these studies were reattacked as "secret science" and unsuitable for public policy. In 2013, a House panel of the U.S. Congress issued a subpoena demanding the "secret science" data of the Harvard Six-Cities and ACS CPS-II cohort studies. (Accusations of "secret science" and related issues will be addressed in chapter 5.) Nevertheless, this relatively calm twelve-year period allowed research consolidation and expansion of the Harvard Six-Cities and ACS studies—resulting in the publication of multiple extended analyses of both cohorts.

Shortly after the independent reanalysis, Arden and ACS researchers began collaborating with several researchers who had been on the HEI reanalysis team. Arden had been impressed with the innovative work by Dan Krewski, Rick Burnett, and others as part of the HEI reanalysis project. Also, the ACS researchers had completed a much longer mortality follow-up of their cohort. Collaborative extended analyses of the ACS cohort made sense.

Over the next fifteen years, the team conducted collaborative research using the ACS CPS-II cohort that included (1) increased mortality follow-ups from seven to twenty-six years; (2) sophisticated statistical models that improved evaluation of spatial patterns and controlled for individual and ecological variables; (3) improved estimates of pollution exposures, including modeled estimates of $PM_{2.5}$ exposures at geocoded residential addresses throughout the United States; and (4) substantially greater statistical power due to the extended mortality follow-ups of more participants.

Collaborative extended analyses of the ACS cohort confirmed the original findings that $PM_{2.5}$ air pollution was a significant environmental risk factor for all-cause, cardiopulmonary, and lung cancer mortality.[18] The extended work explored the links between $PM_{2.5}$ air pollution and cardiovascular disease[19] and cardiometabolic disorders[20] in greater depth, evaluating the biological pathways of disease (discussed in chapter 10).

For example, Michelle Turner, a key collaborator on several extended analyses, led research on air pollution and lung cancer. Her team's analysis of a sub-cohort of never-smokers strengthened previous findings that $PM_{2.5}$ air pollution contributed to lung cancer mortality risk[21] and studies of interactions between cigarette smoking and $PM_{2.5}$ air pollution.[22]

Michael Jerrett, part of the original HEI reanalysis team, held a PhD in geography from the University of Toronto. His geographic information systems (GIS) expertise allowed the team to use spatial epidemiology: mapping estimated pollution exposures. He also contributed to a collaborative extended analysis of the ACS cohort study. Subsequently, he contributed to a series of analyses on ozone pollution and mortality,[23] mortality effects of short-lived greenhouse pollutants,[24] and air pollution and mortality in California.[25] He geocoded the residences of ACS cohort participants and assigned their estimated air pollution exposures based on exposure models, including those using satellite remote sensing. Various modeling approaches revealed links between particulate matter and death, especially models that used ground-based exposure data.[26]

George Thurston, at the department of environmental medicine at NYU School of Medicine, is a strong proponent of understanding which sources and components of air pollution are most toxic and harmful to health. He led a study showing that ischemic heart disease mortality was strongly associated with $PM_{2.5}$ from fossil fuel combustion, including coal smoke and vehicle emissions.[27]

EXTENDED ANALYSIS OF THE HARVARD SIX-CITIES COHORT

Doug and his colleagues also conducted extended analyses of the Harvard Six-Cities study. The first was led by Francine Laden, a professor at the Harvard School of Public Health, who had not participated in the original analysis. Her team evaluated eight additional years of follow-up, and so had the advantage of more statistical power. An exciting aspect of the extended follow-up period was that air pollution exposures declined substantially over the eight years. Overall, the $PM_{2.5}$ mortality observed in the extended analysis was equivalent to that observed in the original. More interesting, however, was the finding that during the extended follow-up, lower air pollution levels tracked with lower mortality risk. The reduction in mortality risk was most prominent in cities with the greatest decreases in $PM_{2.5}$, especially for cardiorespiratory disease but not lung cancer. These results provided evidence that $PM_{2.5}$-associated mortality was at least partially reversible.[28]

In a related extended analysis, Joel Schwartz led an evaluation of the $PM_{2.5}$-mortality relationship and the timing between exposure and effects.[29] They estimated the relationship to be nearly linear and to extend below the National Ambient Air Quality annual standard of 15 $\mu g/m^3$. They also observed that most effects occurred within two years of exposure.

An additional analysis with even longer follow-up found a near-linear $PM_{2.5}$-mortality response relationship.[30] The results were not sensitive to model specifications. Further, the estimated $PM_{2.5}$-mortality associations did not change much over time, even with low pollution levels and a lower proportion of sulfates.

CONCERNS REGARDING MINORITY REPRESENTATION

The Harvard Six-Cities and the ACS CPS-II cohorts did not fully represent the diverse U.S. population. They did not allow for an evaluation of potential susceptibility or exposure inequities—specifically, the cohorts underrepresented minority ethnic and racial groups. The Harvard Six-Cities cohort was relatively small, resulting in analyses that included only white adults. While the ACS CPS-II cohort was much larger, ACS volunteers recruited the participants. Participants were usually friends, neighbors, or acquaintances of the ACS volunteers, resulting in an under-representation of minority and less affluent participants.

It is crucial to understand if there are substantive differences in health effects or exposures across groups of people with different socioeconomic statuses, education, income, and race or ethnicity. Given the underrepresentation of minority ethnic/racial groups in the Harvard Six-Cities and ACS CPS-II cohorts, disparities in pollution-mortality associations across these groups were not adequately explored. An important finding from the reanalysis and extended analyses of the cohorts was that air pollution was more strongly associated with mortality among those with less education. These results suggest the need for further exploration of disparities related to air pollution health effects and exposure in other cohorts.

CONCLUSION

"Do long-term exposures contribute to greater risk of disease and death?"

Yes. Population-based cross-sectional studies have demonstrated that adjusted mortality rates were associated with long-term average concentrations of fine particulate and sulfate particulate air pollution. The Harvard Six-Cities and the ACS CPS-II cohort studies further demonstrated this pollution-mortality association. Specifically, they observed these results in survival analyses of individuals followed for long periods.

Over the years, the data from these two original cohort studies have been analyzed and reanalyzed in many ways and by various research teams. The studies provided evidence that long-term exposure to $PM_{2.5}$ and related air pollution contributes to the risk of overall mortality, respiratory and cardiovascular mortality, and lung cancer. The results of these studies can and have been replicated by independent and extended analyses.

Important questions remain: Can the studies' results be reproduced in places with cleaner air, such as Canada, or with more polluted air, such as China? And how are the health burdens of air pollution distributed across different subpopulations of people?

5

WERE THE COHORT STUDIES REPRODUCIBLE—OR JUST "SECRET SCIENCE"?

On a raw, chilly day in January 1997, Doug traveled to Washington, DC, to testify before a Senate hearing on air pollution. He walked up Capitol Hill from Union Station and encountered a group of apparent protesters. They were dressed in white lab coats and wearing large eyeglass frames. Curious, Doug approached them and found them holding signs that read "Harvard, Release the Data!" and passing out flyers that read "Show Us the Data." It was a bit jarring when Doug realized this was a protest directed at him and the Harvard Six-Cities Study.

As reported by Elaine Grant, the protesters that Doug encountered were "employed by an industry-backed group called Citizens for a Sound Economy." They were making nebulous accusations that Harvard, and Doug, specifically, were inexplicably hiding "secret data."[1] They wielded public relations weapons and tactics initially developed to fight against tobacco marketing restrictions. Doug was somewhat amused by the lab-coat-wearing protesters, but also a bit unsettled as he gave his testimony that day (without further incident).

Doug did not realize, however, that this was the opening salvo in a long-term attack on air pollution epidemiology. Ultimately, it was an attack on the credibility of science in general and would define his career path for the next couple of decades.

As discussed in the previous chapter, Doug, Arden, and colleagues supported efforts to independently reanalyze, validate, and replicate the Harvard Six-Cities and ACS CPS-II studies. Not only had they shared the data, but they had met ethical and legal obligations to protect research participants' confidential information.

After the independent HEI reanalysis results were reported, accusations of "secret science" subsided for a time. However, as we will discuss later in this chapter, rancorous debates over "secret science" erupted again in the early 2010s. The researchers' efforts to find the appropriate balance between adequate review and replication and protecting private data clashed with their adversaries' unabashed efforts to dismiss or nullify scientific evidence related to the health effects of air pollution.

CONTRARIAN USE OF ACS COHORT DATA

Thoughtful critiques of research and its interpretation are helpful and indeed necessary to scientific advances. Critiques and concerns about research results have played an essential role in studying the health effects of air pollution, as discussed throughout this book.

However, at some point in scientific inquiry, the evidence becomes overwhelming. When the evidence has been replicated and validated many times, the scientific community typically reaches a consensus. Unfortunately, that does not always stop people from criticism; in this case, a series of studies were conducted to challenge the ACS cohort analyses. These studies were criticized for their methods—and, in some cases, for their funding sources.

For example, a 2003 study[2] used data from a California subset of participants from the first ACS cancer prevention study (CPS-I) who had been enrolled in 1959. This ACS CPS-I study was the predecessor to the ACS CPS-II study used in the air pollution studies discussed in chapter 4. The 2003 study reported that secondhand tobacco smoke was *not* significantly associated with coronary heart disease, respiratory disease, or lung cancer—results inconsistent with other studies. Perhaps its radically different findings were due to the study's use of only a small subset of the full ACS CPS-I cohort or its unreliable measures of secondhand smoke exposure versus non-exposure. Or perhaps tobacco industry funding had some influence on the reported results.

Issues related to industry funding can be problematic. Although industry funding is not necessarily corrupt, when an industry has a vested interest in the research they are funding, it may create a conflict of interest. For example, the 2003 secondhand cigarette smoke study noted above was cited as one of the tobacco-funded studies used to support the industry's deceptive position on passive smoking in Judge Gladys Kessler's 2006 landmark verdict against major U.S. tobacco companies.[3] And a paper on tobacco industry efforts to undermine data on secondhand smoke noted:

> An analysis of tobacco industry documents revealed that the *British Medical Journal* financial disclosure requirement was not adequate to give readers and reviewers an appreciation for the authors' long-standing relationships with the tobacco industry and the fact that the study was a "special project" funded by industry lawyers and executives outside the peer review process.[4]

In 2005, a study by the same investigator used the same California sub-cohort of ACS CPS-I to look at $PM_{2.5}$ air pollution and mortality among older Californians (mean age of sixty-five years in 1973).[5] The analysis was restricted to only eleven counties with county-level estimates of air pollution. The study found a small but significant mortality association with $PM_{2.5}$ for an initial period (1973–1982) but not for a subsequent period (1983–2002). The paper concluded that there may have been a small pollution effect before 1983, but there was no current effect. A critique of this study[6] described its substantive design and exposure measurement problems, concluding that the study added little understanding of the effects of air pollution.

In 2017, another study by the same author used only a six-year follow-up of ACS CPS-II data.[7] This study reported only a small $PM_{2.5}$-mortality association that was not statistically significant. Unlike the studies reported in chapter 4, this 2017 analysis was not conducted or reported in collaboration with ACS researchers.

The original collaborative research team who had studied the ACS CPS-II cohort published a critical response to this analysis.[8] They noted that the study, even though it was published twenty-two years after the original, could not be considered an extended or advanced analysis of the ACS CPS-II cohort. It had the shortest follow-up, the fewest participants, and the fewest deaths in all the ACS analyses. The critique stated that the study did not use suitable methods for linking participants nor state-of-the-art modeling techniques to estimate exposure.

There have, however, been many constructive research efforts to evaluate the reproducibility of linking cohort studies with data on air pollution and mortality risk, including an analysis of the Women's Health Initiative study data.

ANOTHER COHORT: THE WOMEN'S HEALTH INITIATIVE

Postmenopausal women are at increased risk for several common causes of disease and death, including cardiovascular disease and cancer. Between 1993 and 1998, about 94,000 postmenopausal women across forty clinical centers throughout the United States enrolled in the Women's Health Initiative (WHI) Observational Study.[9] The study aimed to investigate strategies to prevent and control common causes of disease and death among postmenopausal women.

A famous component of the WHI was a randomized controlled study of hormone replacement therapy (HRT) in a subset of WHI postmenopausal women. The study investigated the uncertain notion that HRT increased the risk of breast cancer but protected against coronary heart disease. The trial was stopped early, after only an average of 5.2 years of follow-up. Why was it stopped so soon? The early results indicated that HRT had an unacceptably high risk of breast cancer. Further, and surprisingly, HRT also significantly elevated risks for coronary heart disease, stroke, and deep vein thrombosis or pulmonary embolism. The publication of these results[10] triggered a sharp drop in the use of HRT[11] and consternation among doctors and patients.

The use of the ACS CPS-II cohort to examine the health effects of particulate air pollution revealed the value of adding air pollution exposure data to existing health cohort studies that were not originally designed to study air pollution. Could the WHI cohort also be used to study air pollution and cardiovascular disease?

Research collaborators from the University of Washington, including a key WHI investigator (Garnet Anderson), linked address data from participants in the WHI with available air pollution data. They explored exposure to $PM_{2.5}$ air pollution and the risk of cardiovascular disease in approximately 65,000 postmenopausal women without previous cardiovascular disease. They calculated subjects' exposures to air pollution based on the average concentrations of $PM_{2.5}$ at the closest air pollution monitor to their homes in 2000. They estimated relative risks with statistical models that allowed them to adjust for influential factors such as age, race or ethnic group, smoking status, educational level, household income, and body mass index.

In February 2007, the intriguing results of this analysis were published.[12] Exposure to $PM_{2.5}$ was associated with elevated risks for first cardiovascular events and cardiovascular disease deaths. However, the results of this WHI study were notably different from those observed in the Harvard Six-Cities study or the ACS CPS-II studies. The air pollution–related elevated risks for cardiovascular mortality in these postmenopausal women were inexplicably larger.

OTHER U.S. SUBPOPULATION STUDIES

The WHI study is just one of many studies that subsequently linked air pollution measurements with data from available cohorts and evaluated the death risks of air pollution. Like the WHI study focused on postmenopausal women, many of these studies focused on specific subpopulations. Examples include cohorts of U.S. nonsmoking Adventists, U.S. nurses, U.S. male health professionals, hypertensive U.S. veterans, California teachers, U.S. agriculture workers, and mature adults aged fifty to seventy years enrolled in the NIH-AARP Diet and Health study.[13]

Several studies examined the mortality effects of air pollution on subpopulations of persons with specific health conditions. For example, Arden and colleagues used data for millions of cancer patients and survivors from the U.S. Cancer Surveillance, Epidemiology, and End Results (SEER) program from 2000 to 2016. Of those who died in this cohort, 26 percent died of noncancer causes, primarily cardiovascular and respiratory disease. Fine particulate matter air pollution was associated with an elevated risk of cardiovascular and respiratory disease death for cancer patients and survivors.[14]

In a study of adults with chronic obstructive pulmonary disease (COPD), long-term exposure to $PM_{2.5}$ air pollution was also associated with an elevated risk of cardiovascular mortality.[15]

EUROPEAN STUDIES

European investigators applied similar approaches to multiple existing cohort studies to confirm that long-term air pollution exposure was linked to mortality risk. These include a study from Italy, two from the Netherlands, and one each from England, France, and Denmark.[16]

Perhaps the most challenging study of long-term air pollution exposure and mortality risk was the European Study of Cohorts for Air Pollution Effects (ESCAPE). Its objective was to use information from approximately thirty-five existing cohorts from countries and research teams throughout Europe. Researchers needed to develop standard protocols for measuring, validating, and assigning long-term exposure to air pollution to participants in the cohorts. Researchers also needed to create standard protocols for cohort-specific statistical analyses that could pool the results and develop effect estimates of air pollution. Such a study could be a collaborative and logistic nightmare and would challenge any research coordinator.

Bert Brunekreef is among the world's most respected and productive environmental epidemiologists. He received his PhD from Wageningen University and was a postdoctoral fellow with Doug at the Harvard School of Public Health. Bert is a professor at the Institute for Risk Assessment Sciences at Utrecht University, the Netherlands. He has a polite, gracious personality—and he works well with and is generous to collaborators. Bert had the ideal temperament and skill set to coordinate the highly complex and challenging ESCAPE study. Bert and other highly skilled researchers (including Rob Beelen, Gerard Hoek, Goran Pershagen, Nino Kunzli, Annette Peters, Paolo Vineis, and dozens of others) conducted ESCAPE successfully.

So, what were the findings of the ESCAPE study? "Long-term exposure to fine particulate matter air pollution was associated with natural-cause mortality, even within concentration ranges well below the present European annual mean limit value."[17] Bert and colleagues later conducted other extended analyses of European cohorts. One study (ELAPSE) focused on pollution-mortality relationships at low levels of exposure.[18] Again, air pollution, especially $PM_{2.5}$, was associated with an increased mortality risk.

CANADIAN STUDIES

It's a bit inexplicable. Canada has some of the cleanest air in the world, and yet it has produced remarkably prolific and influential researchers who study the effects of polluted air. David Bates, the young physician in London during the 1952 London smog episode, moved to McGill University in Montreal, Canada, in 1956. In 1972, he moved to Vancouver to become the dean of the faculty of medicine at the University of British Columbia. He was an inspiring pioneer in both lung physiology studies and epidemiological research on air pollution and health. Other outstanding Canadian investigators who were members of the teams that reanalyzed the Harvard Six-Cities or ACS CPS-II cohorts have made many vital contributions to air pollution and health research.

Canada also has provided some of the largest and most comprehensive cohorts that can be linked to state-of-the-art estimates of air pollution exposure—even at very low exposures. These cohorts include millions of participants who completed the long-form census of the Canadian Census Health & Environment Cohorts (CanCHECs) for 1991, 1996, and 2001 and about half a million participants in the Canadian Community Health Survey (CCHS) between 2001 and 2012. Analyses of these Canadian cohorts have consistently found that long-term exposure to $PM_{2.5}$ air pollution was associated with mortality risk.[19]

Michael Brauer is another remarkable Canadian air pollution researcher trained at the Harvard School of Public Health and now a professor at the University of British Columbia. He led a research team to evaluate the morbidity and mortality effects of long-term exposure to low levels of air pollution in Canada. The study had a clever acronym, MAPLE (Mortality-Air Pollution Associations in Low-Exposure Environments). Brauer's team pooled data from approximately eight million subjects in the various Canadian cohorts. Even in Canada, with relatively low levels of air pollution, $PM_{2.5}$ and related air pollution was associated with elevated mortality risk.[20]

CHINESE AND GLOBAL COHORTS

It is an obvious question: If long-term exposure to $PM_{2.5}$ air pollution is associated with mortality risk in relatively low-pollution countries, do we see comparable associations in highly polluted countries such as India or China? Several extensive studies indicate that the answer is "Yes."

One Hong Kong study used satellite-based estimates of $PM_{2.5}$ exposure linked to approximately 67,000 older adults from eighteen Health Centres of the Hong Kong Department of Health. The study evaluated the efficacy of using satellite data to estimate pollution exposure. It found substantial and statistically significant associations between estimates of $PM_{2.5}$ air pollution and all-cause and cardiovascular mortality.[21]

Three large cohort studies have been reported from mainland China. The first included approximately 190,000 men, at least forty years old, from forty-five areas of China. Long-term $PM_{2.5}$ exposure was adversely associated with all-cause, cardiovascular, and lung cancer mortality.[22] The second study included over 13,000 Chinese adults aged sixty-five or older. This study found that $PM_{2.5}$ air pollution was adversely associated with all-cause mortality.[23] The third study used approximately 117,000 adults eighteen years of age and older pooled from four prospective Chinese cohorts. Again, $PM_{2.5}$ air pollution was adversely associated with mortality, including non-accidental and cardiovascular or "cardiometabolic" mortality.[24]

In November 2017, an international study reported remarkable and provocative results. The Prospective Urban Rural Epidemiology (PURE) study was a large cohort study of persons thirty-five to seventy years of age from many countries on five continents. The results suggested that high-carbohydrate diets were associated with a higher mortality risk, but total fats and some specific types of fats were associated with lower mortality risk.[25] A team of researchers, led by Michael Brauer and Perry Hystad, recognized that this large multi-city cohort included locations with high levels of air pollution (such as India and China). Thus, it could be a valuable resource for studying air pollution effects. They linked air pollution data to this cohort's 747 urban and rural communities. Using traditional survival analyses and adjusting for individual and community characteristics, they found that $PM_{2.5}$ air pollution was not associated with all-cause mortality but *was* adversely associated with increased risk of cardiovascular disease events (especially stroke) and death.[26]

It's unclear how many more cohort studies of air pollution and mortality are needed. As briefly outlined in this chapter, growing numbers of cohort studies worldwide have reported links between $PM_{2.5}$ air pollution and mortality. It seems evident that the results of these studies should inform air quality standards and public policy efforts to address air pollution. There is, however, a concerted effort to dismiss these studies.

"SECRET SCIENCE" SAGA

Chapter 4 noted a twelve-year lull in the controversies focused on the Harvard Six-Cities, ACS CPS-II, and subsequent cohort studies. During this time, Arden and Doug accepted an invitation to write a critical review of the literature on the health effects of $PM_{2.5}$ air pollution.[27] About ten years earlier, Sverre Vedal, a pulmonologist and respiratory epidemiologist, wrote an insightful review of previous literature. His critical review was written during the polarized debate over air pollution science and the proposed new $PM_{2.5}$ standards. Vedal aptly titled his review "Ambient Particles and Health: Lines That Divide."[28]

By 2006, when Arden and Doug's new review was published, much had changed. The title of the review was "Health Effects of Fine Particulate Air Pollution: Lines That Connect."[29] This title suggested that the science of $PM_{2.5}$ pollution was advancing and, in some ways, consolidating. Doug and Arden knew that the various divisive issues, controversies, and contentious debates about air pollution science and public policy had not been fully resolved. However, their "Lines That Connect" review was optimistic. Maybe some of the contentious rancor that had been part of earlier research was giving way to more concerted and cooperative efforts to understand and address the adverse health effects of air pollution. In some ways, they were right to be optimistic. Many research teams and projects were advancing the knowledge of the health effects of air pollution. Further, serious public policy efforts were being made to address this issue.

In other ways, Doug and Arden were wrong. As noted earlier, accusations of "secret science," often carefully orchestrated, occurred after the Harvard Six-Cities and ACS-CPS-II cohort studies were published. In the early 2010s, a few years after the "Lines That Connect" review, insinuations increased that fundamental air pollution studies were "secret science" and not suitable to inform public policy. The "secret science" argument was made about not only the Harvard Six-Cities and ACS CPS-II cohort studies but also other cohort and clinical studies that collected and used private and confidential medical data.

Those calling for the data to be released chose to ignore or minimize the ethical and legal restrictions against sharing people's private health information. When individuals are recruited and enrolled to participate as research participants in a cohort study, private information of all sorts is obtained. For example, individual information regarding birth dates, death dates, and cause of death is fundamental to statistical survival analyses. Place of residence is needed to assign estimates of pollution exposure. Information including age, sex, race/ethnicity, smoking history, health conditions and existing disease, medication use, education levels, marital status, income, occupation, and other variables are also crucial to the analysis. When research subjects are recruited and enrolled, they are guaranteed that their personal information will be kept confidential.

In addition, guarantees of confidentiality are required by institutional review boards that approve research using human subjects, as well as agencies providing vital medical records and mortality data. Doug, Arden, and their collaborators support the principles of data accessibility and transparency in conducting and reporting their research. They also understand that conscientious, trustworthy researchers respect and adhere to ethical and legal obligations to protect the confidentiality of research subjects. Despite these confidentiality obligations, the "secret science" debate seemed to be about shifting the public policy focus to be more about process rather than a critical evaluation of the scientific evidence.

Complaints about "secret science" were more than just idle sniping. In May 2013, Mark Drajem, a reporter from *Bloomberg News*, contacted Doug and Arden separately and asked them to respond to a news article he was writing. At the time, Gina McCarthy was President Barack Obama's nominee for U.S. EPA administrator, but her confirmation was being stalled by U.S. senators who demanded the release of the Harvard Six-Cities and ACS CPS-II data.[30] Mark asked Doug and Arden, as coauthors of these studies, for their responses.

Until receiving this call, Arden and Doug were not fully aware of the renewed political wrangling regarding the Harvard Six-Cities and the ACS CPS-II cohort studies. However, they both reminded this reporter (and subsequently other reporters) that the data from these studies had already been made available—with the protection of private data—for independent reanalysis over a decade ago. Independent and extended analyses, as well as various new analyses with other cohorts, were conducted.

The revived "secret science" controversy regarding these studies seemed contrived. But contrived or not, it was being used to delay the confirmation of Gina McCarthy as administrator of the EPA.[31] McCarthy's nomination was eventually confirmed, but a new round of political and public policy controversy regarding key studies of air pollution and accusations of "secret science" had been stoked.

In 2013, Senator David Vitter and other U.S. senators requested that the EPA release all the data files for the Harvard Six-Cities and the ACS CPS-II cohort studies, including the "coding of personal health information."[32] On August 1, 2013, the U.S. House of Representatives Committee on Science, Space, and Technology issued a rare subpoena to the EPA to provide access to research data.

The EPA then requested data from Doug, Arden, and other researchers. Requests included data, protocols, methods, and related information from the original Harvard Six-Cities and ACS CPS-II cohort studies, the independent reanalysis, and the expanded analyses. They also requested data for another study of changes in air pollution and U.S. life expectancies (discussed in chapter 7), including data on air pollution exposure measurements, protocols, and methods. The life expectancy study had no individual confidential information, and the researchers provided all data as requested. But Harvard and ACS researchers remained committed to meeting their ethical and legal obligations to protect research participants' personal and confidential information. No private, confidential data on research participants were provided.

Next, Lamar Smith, the chair of the U.S. House of Representatives Committee on Science, Space, and Technology, sponsored federal legislation prohibiting the EPA from using studies for any policy purposes unless the data and related information were publicly available. The proposed legislation included the "Secret Science Reform" Acts of 2014 and 2015 and the HONEST (Honest and Open New EPA Science Treatment) Act of 2017. When even these legislative efforts failed to pass, an alternative approach to prohibit the use of these studies was attempted. This approach was a rule proposed by the EPA administrators during the Donald Trump administration. The proposed rule, euphemistically called "Strengthening Transparency in Regulatory Science," forbade using any study that did not make its underlying data publicly available and validated by EPA review. Critics of the rule quickly noted that this was another effort to "weaken air pollution regulations by barring key studies."[33]

Why the effort to block the use of cohort studies? As evidence grew about the link between air pollution, death, and health problems, "the EPA came under increasing pressure to strengthen its air quality standards."[34] Industries contributing to air pollution knew that reducing their emissions was costly. Hence, they pressured sympathetic politicians to support legislation trying to dismiss scientific research and government regulations that address air pollution.

There is broad support for data sharing and transparency in science. Yet major medical and scientific organizations soundly denounced the "Strengthening Transparency in Regulatory Science," or "secret science," rule. Putting researchers in an impossible legal and ethical position effectively meant that research using personal health information could never be used to inform public policy. The American Association for the Advancement of Science stated that the rule "weakens the use of science in policy-making" and is "a deliberate attempt to exclude scientific evidence from the policy-making process."[35] The editors of major scientific and medical journals, including *Science, Nature, Public Library of Science (PLOS) Journals, Proceedings of the National Academy of Sciences (PNAS), Cell,* and the *Lancet,* issued an extraordinary joint statement on the proposed rule.[36] The editors expressed concern that this rule "would be used as a mechanism for suppressing the use of relevant scientific evidence in policy-making, including public health regulations." They emphasized their support for sharing data and transparency in science but also clearly recognized "the validity of scientific study, that for confidentiality reasons, cannot indiscriminately share absolutely all data."[37]

So, what happened to these extended efforts to prohibit using crucial air pollution studies to inform environmental and public health policy? The congressional subpoena did not compel researchers to violate confidentiality obligations to research participants. The "secret science" and HONEST acts never passed. Under the EPA leadership of Scott Pruitt and Andrew Wheeler, and with considerable opposition, the "Strengthening Transparency in Regulatory Science Rule" was fast-tracked and made effective on January 6, 2021, just before President-elect Joe Biden took office.

However, on January 27, 2021, Brian Morris, Chief District Judge, United States District Court, ruled that the fast-tracked action violated federal law. On February 1, 2021, the court vacated and remanded the rule to the EPA.[38] On May 26, 2021, the EPA implemented the court's decision, overturned the rule, and reaffirmed the EPA's commitment to using the best available science, including studies that included some confidential data.

The "secret science" accusations, however spurious, had an impact. They shifted public policy focus to the scientific and public process rather than an attentive evaluation of scientific evidence. But there was a positive consequence of the "secret science" debate. It provided extra motivation for at least two substantiative research efforts based on *available public data.*

U.S. MEDICARE COHORT STUDY

Given the "secret science" debate, it seemed imperative to conduct cohort studies of long-term exposure to air pollution using data that were already public. Francesca Domenici and Joel Schwartz assembled a stellar research team (including Qian Di, Petros Koutrakis, Antonella Zanobeti, and others) for such an analysis. Later, Francesca noted that one motivation for the study was the opportunity to use publicly available data. She stated, "This is a very highly contentious political climate, and we are taking the extra step to be as transparent as we can be."[39]

The analysis was so large that it put the team's big data management, statistical tools, and computer facilities to the test. The team constructed a cohort of all Medicare beneficiaries in the continental United States from 2000 to 2012. This cohort included approximately 61 million persons with nearly half a billion person-years of follow-up. The team linked estimates of $PM_{2.5}$ and ozone air pollution exposures with zip codes of residence and conducted a statistical analysis. As illustrated in figure 5.1, they found that $PM_{2.5}$ air pollution was associated with elevated mortality risk. The researchers also observed a smaller but statistically

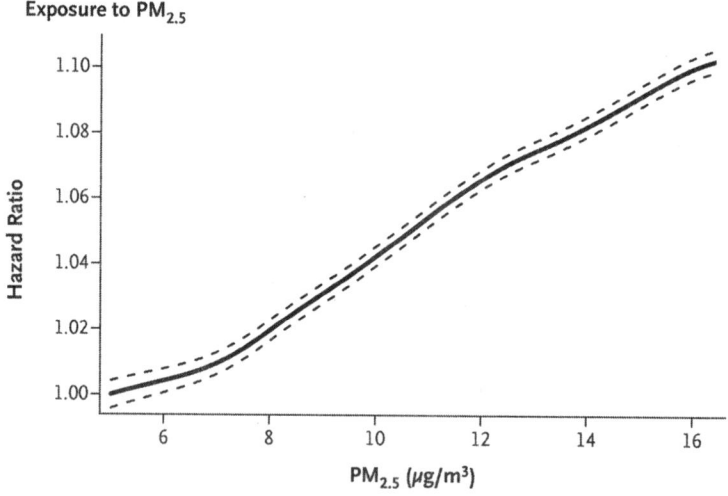

Figure 5.1

Concentration-response relationship between mortality risk and $PM_{2.5}$ from the U.S. Medicare population. *Source:* Qian Di, Yan Wang, Antonella Zanobetti, Yun Wang, Petros Koutrakis, Christine Choirat, Francesca Dominici, and Joel D. Schwartz. "Air Pollution and Mortality in the Medicare Population," *New England Journal of Medicine* 376, no. 26 (2017): 2513–2522. Reprinted with permission from Massachusetts Medical Society.

significant and independent association between ozone air pollution and mortality risk.[40]

The U.S. Medicare cohort study had some drawbacks, including limited availability of some individual-level variables. And this Medicare cohort is obviously skewed toward a relatively older population. Approximately two-thirds of all deaths in the United States occur in this age group. The study statistically adjusted for key demographic characteristics, Medicaid eligibility, and area-level differences. Given the massive size of the cohort, the accompanying statistical power, and the quality analyses, this study was impressive.

Importantly, in the context of the "secret science" arguments, the U.S. Medicare study used publicly available beneficiary file data from the Centers for Medicare and Medicaid Services. However, these data are de-identified, with restricted use. The restrictions protect patient privacy and medical record information consistent with federal Health Insurance Portability and Accountability Act (HIPAA) privacy rules. Additionally, there are other reasonable safeguards to protect patient privacy.

CACES STUDIES USING PUBLICLY AVAILABLE DATA

In 2015, Arden joined a research team as part of the Center for Air, Climate, and Energy Solutions (CACES). This center was codirected by Allen Robinson, a professor of engineering and public policy at Carnegie Mellon University, and Julian Marshall, a civil and environmental engineering professor at the University of Washington. Many of the researchers on this team were engineers. Much of the work conducted by CACES involved advanced methods for measuring and modeling air pollution exposures and evaluating public policy scenarios and outcomes. One subproject, however, involved epidemiologic research. The team for this subproject took estimates of air pollution exposures generated by CACES for use in epidemiological studies of air pollution and mortality.

Rick Burnett persuaded Arden to be a co-investigator on this subproject. What was Rick's enticement? He argued they could construct a study cohort using U.S. National Health Interview Survey (NHIS) data linked with mortality data. This cohort would be large. It would represent the U.S. population, and the team could connect the cohort with air pollution estimates generated by the engineers and atmospheric scientists from CACES. The air pollution estimates generated from CACES would be made publicly available. The NHIS data is a public data set. Jennifer Parker and colleagues at the National Center for Health Statistics (NCHS) had already conducted an excellent analysis of air pollution and heart disease using a shorter follow-up from NHIS data.[41]

However, given the "secret science" controversy, Rick argued that the team should undertake an expanded study. He contended that analyzing air pollution and mortality using this large extended representative cohort constructed from public data would be an important contribution.

Rick was correct. The study's results further confirmed that long-term exposure to $PM_{2.5}$ air pollution was associated with an increased risk of all-cause mortality.[42] The estimated relationship between average long-term $PM_{2.5}$ concentrations and all-cause mortality was similar to the Medicare Study, as illustrated in figure 5.1. The NHIS cohort also allowed analysis by specific causes of death. $PM_{2.5}$ air pollution was most strongly associated with cardiovascular, cardiopulmonary, and lung cancer mortality. Figure 5.2 illustrates the estimated relationship between average long-term $PM_{2.5}$ concentrations and cardiopulmonary mortality. This study was not as large as the Medicare study discussed earlier. But it was more representative of U.S. adults of all ages and could control for more individual risk factors, including smoking.

The NHIS public data also had some use restrictions. The team linked air pollution exposure data to the NHIS study participants at the census tract level, which required information on participants' census tract of residency. The presence of location information made participant confidentiality more challenging. Even without a name attached, a home address could potentially be used to identify subjects. Any use of the data files had to comply strictly with data management and analytic procedures, so the subjects remained de-identified.

To further protect the confidentiality of the study participants, the research team was required to conduct the analyses of the NHIS data at secured, restricted-use data centers using secured computers provided at the centers. Most of the analyses were conducted at the NCHS Research Data Center in Hyattsville, Maryland. An unexpected COVID-19–related challenge occurred while conducting the analyses on the NHIS data. But

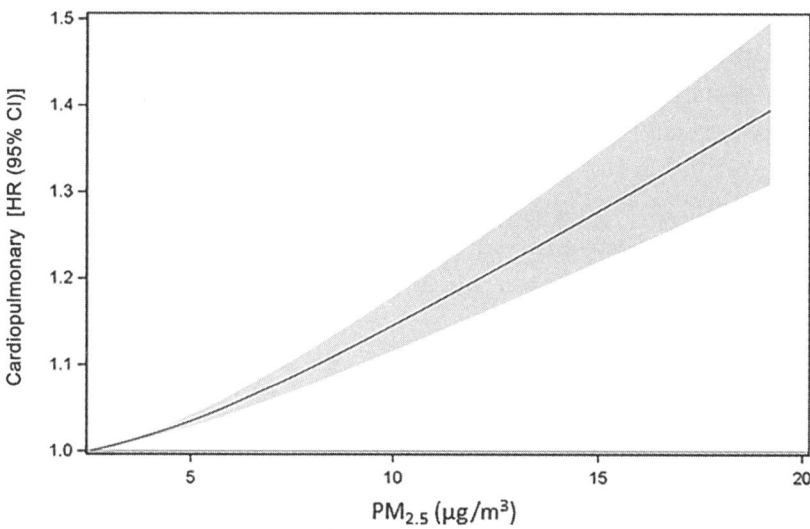

Figure 5.2
Concentration-response relationship between cardiopulmonary mortality risk and PM$_{2.5}$ from the NHIS cohort study. *Source:* C. Arden Pope, Jacob S. Lefler, Majid Ezzati, Joshua D. Higbee, Julian D. Marshall, Sun-Young Kim, Matthew Bechle, et al., "Mortality Risk and Fine Particulate Air Pollution in a Large, Representative Cohort of U.S. Adults," *Environmental Health Perspectives* 127, no. 7 (2019): 077007.

this challenge resulted in the fortuitous use of another publicly available research data source and insights into the potential effects of air pollution on cancer patients.

During part of the COVID-19 pandemic period (2020–2021), travel restrictions and the closure of the secured research data centers temporarily sidelined the members of the research team from analyzing the NHIS data. (They could and did resume the work after the centers reopened.) But what could be done in the interim? Nathan Coleman, one of Arden's research assistants, had an idea. While working the previous summer as an intern at a consulting firm, Nathan had become familiar with the SEER cancer registry data. This data source included relevant information on millions of cancer patients compiled from the U.S. National Cancer Institute's Surveillance, Epidemiology, and End Results (SEER) program. The use of these data required signing a SEER data-use agreement, which included committing to protect patient confidentiality. The data analyses, however, could be conducted using the research team's own computers and at their own university research facilities—even while taking precautions during the COVID-19 pandemic.

So, how did Nathan's idea work out? How did the research team use these SEER data and what did they learn? First, the cancer registry data were linked with air pollution exposure data generated by other CACES collaborators. They observed that $PM_{2.5}$ air pollution was associated with cancer incidence, especially lung cancer.[43] Second, the SEER cancer registry data was used to construct a cohort of over 5.5 million adult cancer patients for the years 2000–2016. Of the patients who died, 26 percent died of noncancer causes, mostly from cardiovascular and respiratory diseases. Careful analyses of the SEER cohort revealed that cancer patients were especially vulnerable to the effects of air pollution contributing to cardiopulmonary disease. Elevated exposures to $PM_{2.5}$ air pollution increased the risk of dying of cardiovascular and respiratory disease (approximately 25 percent increase in risk per 10 $\mu g/m^3$ increase in $PM_{2.5}$ exposure). The increased risk was relatively high for those who received chemotherapy or radiation treatments.[44]

FORESTS OF EVIDENCE

Chapters 4 and 5 have discussed many individual cohort studies of long-term $PM_{2.5}$ exposures and mortality risk. A forest plot as presented in figure 5.3 helps summarize these studies in a quick and easy visual way.[45] We generated this forest plot to illustrate how the results from the early cohorts have been reproduced over and over.

Figure 5.3 zooms out to show the bigger picture—the forest versus individual trees. This figure shows standardized estimates of the $PM_{2.5}$-mortality associations for all-cause and cardiopulmonary mortality from major studies and allows an evaluation of the overall pattern of associations. The standardized estimates are mortality hazard ratios (HR) with 95 percent confidence intervals for a difference in $PM_{2.5}$ of 10 $\mu g/m^3$. An HR of 1.0 indicates that there is no $PM_{2.5}$-mortality association. HRs greater than 1 indicate an adverse $PM_{2.5}$-mortality association. For example, an HR of 1.08 indicates that a 10 $\mu g/m^3$ increase in $PM_{2.5}$ is associated with an 8 percent increase in mortality risk.

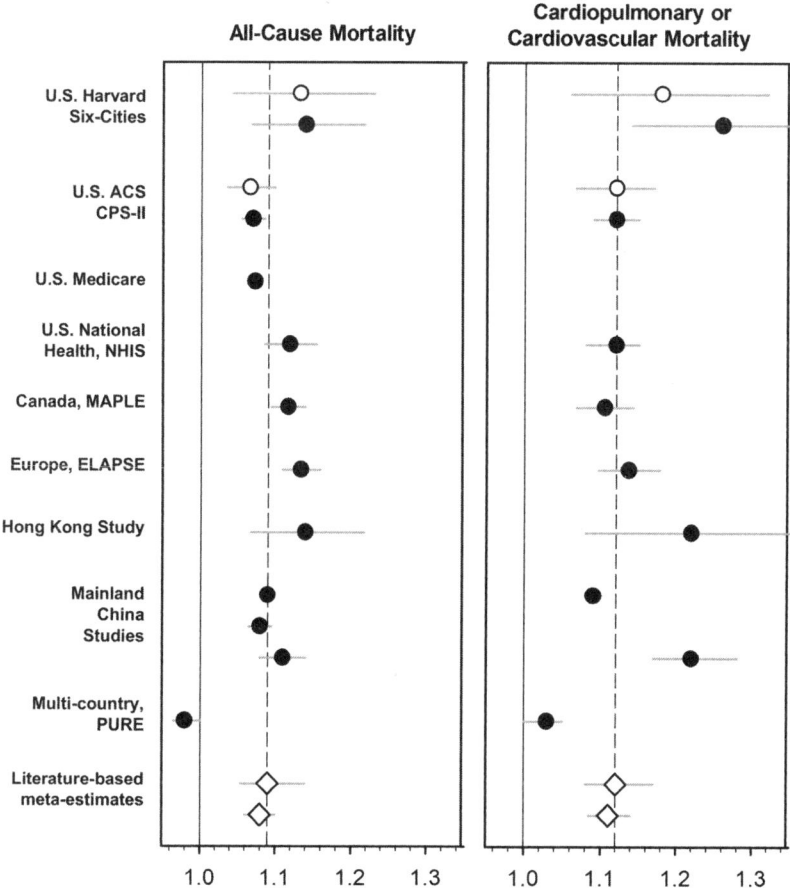

Figure 5.3

A forest plot presenting standardized mortality hazard ratios across key studies. Estimates are per 10 μg/m³ PM$_{2.5}$, For the Harvard Six-Cities and ACS CPS-II studies (open circles are original estimates, closed circles are estimates from latest extended analysis) and for the key largest/latest studies. Diamonds present literature-based summary or meta estimates of overall cohort studies. *Sources:* See note 45.

So, what appears in the forest plot (figure 5.3), and why is it so remarkable? The first estimates are from the Harvard Six-Cities and the ACS CPS-II cohorts. Open circles are original estimates; closed circles are estimates from the latest extended analysis. Estimates for other large studies include the U.S. Medicare study, the U.S. NHIS study, the Canada MAPLE study, the Europe ELAPSE study, the studies from Hong Kong and mainland China, and the multi-country (PURE) study. Recently published meta-estimates, which are summary estimates based on careful, systematic reviews of the broader literature, are also presented.[46] The forest plot does not present every single study. The plot would be too busy and messy (only the biggest and most robust trees are shown).

What is the big picture? Are the early cohort studies reproducible? Yes. Further, an analysis of the overall evidence, as presented in figure 5.3, indicates most of the studies are finding similar results—with estimates close to the dashed line that represents a central or average estimate. The results generally indicate that a 10 μg/m^3 increase in PM$_{2.5}$ exposure contributes to a 6–9 percent increase in all-cause mortality, a 9–14 percent increase in cardiovascular or cardiopulmonary mortality, and a 7–16 percent increase in lung cancer mortality.

All the studies don't get exactly the same result. Given the differences in the studies, getting precisely the same result would be suspicious. But the studies are remarkably consistent.

CONCLUSION

"Were the cohort studies reproducible—or just 'secret science'?"

The studies were reproducible, not just in the United States but globally. Studies by research teams worldwide reproduced evidence that long-term exposure to PM$_{2.5}$ air pollution contributes to the risk of mortality from all-cause, cardiopulmonary, and lung cancer disease.

And what about the "secret science" accusations? They have little merit. The cohort studies' protocols, methods, and results are openly published in high-quality, peer-reviewed scientific journals. The Harvard Six-Cities and the ACS CPS-II cohort studies provided data access for audits, independent analyses, and extended analyses. Research teams conducted the studies in collaborative efforts to expand scientific understanding of the effects of air pollution. Notably, the research teams conducted and reported the research in such a way as to protect the privacy and confidentiality of research participants. The protection of privacy and confidentiality is required even when using publicly available data such as U.S. Medicare data and U.S. National Health Interview Survey data.

"Secret science" accusations, for a time, obscured fundamental ethical and legal obligations to protect the privacy and confidentiality of research subjects. They created a deceptive smog of demands for "full transparency" that would breach those obligations. Their goal was not transparency but to prevent the use of major studies that established a link between air pollution and adverse health effects. The "secret science" accusations shifted the focus from evidence to process, at least for a while. The evidence, however, is well documented and compelling.

6

ENVIRONMENTAL JUSTICE AND AIR POLLUTION: WHO PAYS?

The evidence presented in previous chapters indicates that air pollution broadly affects the health of all exposed humans—adults and children of various racial-ethnic groups, and populations worldwide. But some are more adversely affected by air pollution than others.

A special report in *Scientific American* tells the story of an African American family, the Clarks, in California that grew up within a mile of a refinery. The authors, Jane Kay and Cheryl Katz, described how Henry Clark, as a child, could see dark plumes of smoke pouring out of the refinery's stacks. "Sometimes I'd find the air so foul, I'd have to grab my nose and run back into the house until it cleared up,"[1] Clark said. Kay and Katz wrote:

During World War II, African Americans like Clark's family moved to homes in the shadow of this refinery because they had nowhere else to go. Coming to California looking for opportunity, they quickly learned that white neighborhoods and subdivisions didn't want them.[2]

The Clarks, and other minorities nationwide, experienced a "triple whammy" of race, poverty, and environment, according to Kay and Katz. These forces converged to push minorities into polluted neighborhoods. Kay and Katz added that Black leaders in the Civil Rights Movement called the phenomenon "environmental racism."[3]

There is mixed evidence that specific subgroups of people are more susceptible to the effects of air pollution than others.[4] For example, the vast U.S. Medicaid cohort evaluated multiple subgroups' air pollution and mortality risk. The study found that $PM_{2.5}$ air pollution was more strongly associated with mortality risk for nonwhite populations, especially Hispanic and Black populations. The study also found that $PM_{2.5}$ air pollution was more strongly associated with those with relatively low incomes and who were eligible for Medicaid. The authors of this study noted that their results suggest that adverse health effects of air pollution were "most pronounced among self-identified racial minorities and people with low income."[5]

Analyses of large, representative cohorts constructed from U.S. National Health Interview Survey study data also observed more severe air pollution effects for self-identified Hispanic and Black populations. Yet these differences were not as pronounced and not statistically significant.[6]

Although there is mixed evidence regarding differences in susceptibility to the health effects of similar levels of air pollution, there is a finding regarding environmental inequities that is unambiguous. Substantive differences in air pollution *exposures* exist across different groups of people.

In the United States, there are disparities in exposure to air pollution across ethnic and racial groups, income levels, and education levels. One careful analysis of exposure to $PM_{2.5}$ air pollution across the United States found that racial-ethnic minorities are exposed, on average, to approximately 14 percent higher levels of ambient fine particulate air pollution.[7] Notably, the study's findings indicate that the racial disparities were at least partly independent of income. Christopher Tessum, lead author of the study, stated:

Some assume that when there is a systematic racial-ethnic disparity, such as the one we see here, the underlying cause is a difference in income. Because the data shows that the disparity cross-cuts all income levels, our study reinforces previous findings that race, rather than income, is what truly drives air pollution–exposure disparities.[8]

In 2022, another research team linked demographic and air pollution data across the United States. They found worse $PM_{2.5}$ air pollution in areas with higher than average ethnic-racial minority populations, especially Black, Asian, Hispanic, and Latino populations. Further, they also found that areas with low-income populations were consistently exposed to higher levels of air pollution.[9]

A remarkable 2022 study explored the impacts of historically discriminatory mortgage appraisal practices, sometimes called "redlining." In the 1930s, the United States federally sponsored Home Owners' Loan Corporation (HOLC) constructed maps that graded neighborhoods. Many neighborhoods were assigned the worst grades (redlined) primarily because they included higher black and immigrant populations. These neighborhoods were commonly ineligible for federally backed loans with favorable mortgage terms.

Although "redlining" programs were discontinued in 1968 when the Fair Housing Act banned racial discrimination in housing, racial and socioeconomic disparities persist across these neighborhoods. Air pollution exposures, including $PM_{2.5}$ and NO_2, are consistently and substantially higher in these historically "redlined" neighborhoods.[10]

Earlier studies are consistent with findings about pollution exposure disparities across race and income. A 2015 analysis looked at twenty-two North American studies and found a pattern in which low-income neighborhoods had more exposure to air pollution. The analysis also found that lower educational levels (below a high school diploma) were linked with higher exposure.[11]

Michelle Bell, a professor of environmental health at Yale University, is exploring the relative health impacts of various components of $PM_{2.5}$. She was trained as an environmental engineer at MIT, Stanford, and Johns Hopkins and subsequently worked with the Hopkins team on multiple studies of air pollution and health (see chapter 3). With Keita Ebisu, Michelle found racial and ethnic disparities in air pollution exposure for thirteen of fourteen specific components of $PM_{2.5}$. For example, Hispanics had 152 percent higher exposures for chlorine and 94 percent higher for aluminum than non-Hispanic whites. Further, income and employment status made a difference. Areas with 10 percent more unemployed people were linked with 20 percent more vanadium and 18 percent more elemental carbon exposure.[12]

On the state level, a 2019 California study by the Union of Concerned Scientists found that on average Black and Latino households were exposed to approximately 40 percent higher levels of $PM_{2.5}$ pollution. The study also found differences by income: the lowest-income households in California are exposed to 10 percent more $PM_{2.5}$ than the state average, while the highest-income households have an exposure 13 percent below the state average.[13]

Why are racial minorities and people with low income and education exposed to higher air pollution? No one would willingly choose to live near smelly, dangerous air pollution sources. Yet those with fewer resources or influence are less able to "vote with their feet" and move elsewhere. They are also less likely to successfully contest the placement of pollution sources in their neighborhoods. Air pollution sources such as freeways, power plants, and factories are more commonly located in areas with low real estate values and where people with lower incomes reside.

As noted above, racial and income-related disparities[14] have long affected pollution exposure. The authors of a 2018 study on environmental toxicology stated, "Accounts of the industrial division of labor by race in major U.S. cities document how people of color were restricted to low-wage, hazardous occupations while simultaneously being confined to low-income housing near these industries."[15]

Another issue in exposure disparities is a lack of knowledge of the risks of air pollution. People may know that a smoggy neighborhood is unpleasant but not know how dangerous it is to their health. For example, consider Ella, the little girl in London whose story was presented in chapter 1. Ella's mother, Rosamund, did not learn about the health effects of air pollution until *after* Ella's death. Further, Ella's physicians could not tell Rosamund what was causing her daughter's illness. Ella's family did not have access to healthcare that was sufficiently informed about environmental hazards. As mentioned in chapter 1, after learning that air pollution contributed to the death of her daughter, Rosamund engaged in the fight to increase knowledge and awareness of the risks of air pollution and founded the organization "Clean Air for All."[16]

Ella's case is even more poignant when contrasted with that of Kristina, a child also facing adverse health effects from air pollution. Unlike Rosamund, Kristina's parents had the resources, including money, education, and access to knowledgeable healthcare providers, to help their child.

Within the first few months of Kristina's life in 1987, she had difficulty breathing; her airways would swell shut. She had to sleep sitting up and could not be left unattended at night. And she couldn't go outdoors.

"We went to the family doctor, and he started referring us to special-ists. Three individual doctors were telling us the exact same thing: That she was just sensitive to the horrific air quality in our county. All three of them," said Kristina's mother, Kimberly.[17]

As mentioned in chapter 1, Kristina's parents installed a high-quality air filter in their house and bought a second home near a ski resort, hop-ing for better air quality. But that's not all. Armed with the knowledge that pollution was harming her child's health, Kimberly joined other par-ents to form an advocacy group, the Utah Clean Air Coalition. Kimberly described visiting the Utah State Bureau of Air Quality offices weekly.

"I'd drive up there, and the director, Burnell Cordner, took me under his wing because his mom lived in the valley, and he knew the problems we were having. He was very supportive, and he educated me on reading smoke, the opacity from the stacks, and he was teaching me everything he could. I said, 'I've got to find other women. Other moms like me.' He opened up the drawer, a file of complaints by other moms from Utah Val-ley, and that's how I found Mary and Julie. And we formed this coalition, the three of us together. It was kind of exciting there for a year. It still goes on," she said.[18]

Part of the "excitement," unfortunately, involved threats. People who Kimberly described as "thugs" approached her while she was distributing the Utah Clean Air Coalition's newsletter. She says they told her, "Well, your husband needs to take his job and move it because there's going to be a bomb in your mailbox. Don't start your car."[19]

Even for a family with knowledge and resources, fighting for clean air was difficult and potentially dangerous. For families that don't know the health risks they face, the fight isn't even an option, and it is an uphill battle without resources. The public needs accurate information about environmental risks.

Air pollution exposure disparities are even more pronounced when evaluated across populations globally. Annual average exposure to air pol-lution in the polluted cities of India, China, Nigeria, Bangladesh, Pakistan, and other countries is many times greater than in the United States.[20] On November 5, 2022, the day the authors drafted this paragraph, $PM_{2.5}$ con-centrations in Salt Lake City and Boston ranged between 2 and 10 μg/m^3. In Lahore and New Delhi, two of the most polluted cities in the world,

PM$_{2.5}$ concentrations ranged from 150 to 500 µg/m^3. One news report of air pollution in New Delhi at this time was starkly headlined "Toxic Smog Turns India's Capital into a Gas Chamber,"[21] and another stated: "Delhi's Air 'Branded' Hazardous, Spurs Calls to Close Schools."[22] Notably, the nations listed above with high air pollution concentrations also have far lower per capita incomes than the United States.[23]

Such exposures have consequences. A 2016 global study estimated that 94 percent of deaths due to environmental pollution (all kinds, not just air pollution) were in low- and middle-income nations.[24]

Within highly polluted cities, there are further exposure disparities based on income levels, housing, school, type of transportation, workplaces, and the ability to avoid exposure. A remarkably illustrative project explored differences in exposure to PM$_{2.5}$ pollution between two New Delhi schoolchildren, Monu and Aamya. These children came from families with substantially different incomes, neighborhoods, housing, schools, and transportation.

This project was conducted by the *New York Times* in collaboration with various air pollution researchers and published with interactive graphics and cinematography as a special report titled, "Who Gets to Breathe Clean Air in New Delhi?"[25] The researchers found that children in New Delhi, over a typical day, were exposed to extremely high levels of air pollution. However, the low-income family's child, Monu, experienced approximately *four times* as much pollution as the affluent family's child, Aamya.

The disparity in the children's lung exposure to air pollution is illustrated in figure 6.1, which shows the personal PM$_{2.5}$ air pollution filters. The darkening from PM$_{2.5}$ pollution on both filters indicates that both children had substantial air pollution exposure. However, the darker filter on the left was from Monu, the low-income child who was exposed to substantially more pollution for nearly all of his activities, including studying, eating, playing, and sleeping at home, or while commuting to and attending school. This project may not fully represent the degree of exposure disparities in New Delhi or other highly polluted cities. However, it anecdotally illustrates a stark reality. There are often large inequities in who bears the health burdens of air pollution.

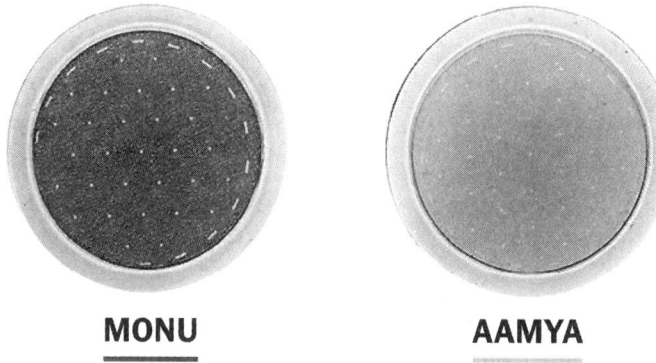

MONU   **AAMYA**

Figure 6.1

Photos of personal $PM_{2.5}$ air pollution filters for a typical day from two children from two different socioeconomic backgrounds in New Delhi, India. Monu's filter, on the left, represents approximately four times as much pollution exposure as Aamya's filter, on the right. *Source:* Adapted from a color photo from Jin Wu, Derek Watkins, Josh Williams, Shalini Venugopal Bhagat, Hari Kumar, and Jeffrey Gettlemans, "Who Gets to Breathe Clean Air in New Delhi?," *New York Times.* December 17, 2020, https://www.nytimes.com /interactive/2020/12/17/world/asia/india-pollution-inequality.html. *Credit:* Leslye Davis /The New York Times/Redux.

CONCLUSION

"Who pays the costs of air pollution?"

Racial and ethnic minorities and people with low socioeconomic status pay a disproportionate burden of the cost of air pollution. Pollution impacts health and quality of life. While air pollution affects everyone exposed, minority and low-income communities around the world are more likely to be exposed to higher concentrations of air pollution.

As discussed earlier in chapter 2, public policy controversies, and even some scientific controversies regarding air pollution and health, are at least in part related to issues of who gets the benefits and *who pays the costs of air pollution.* These disparities play an important role as we seek local and global solutions to control air pollution and improve public health effectively and equitably.

7

DOES REDUCING AIR POLLUTION IMPROVE HEALTH AND REDUCE MORTALITY?

In January 2009, Doug and Arden published research showing that reduced exposure to air pollution reduced mortality risk and increased life expectancy.[1] This research will be discussed later in this chapter.

Shortly after publication, they were amused when comedian Stephen Colbert cleverly mocked their findings on *The Colbert Report*. On this Comedy Central talk show, Colbert took on the persona of a fictional conservative political pundit. In a sketch called "Cheating Death—Lung Health," Colbert wryly joked that the increase in life expectancy would allow watching an extra season of the sometimes-maligned TV series *Ghost Whisperer*. He suggested that "if air with fewer particles will extend your life for five months, logically air with no particles should extend it indefinitely." He then satirically introduced a lifesaving device called "Lifebag" that you place over your head to protect your lungs from harmful airborne particles.[2] Unfortunately, Colbert's satire was accurate in its portrayal of the logical fallacies, if not the content, used by some pundits to discredit science.

Whether young, old, or a comedy writer, one can treat mortality risk glibly, seriously, or comically. Mortality risk is often misunderstood. Isn't our lifetime risk of death 100 percent? Well, yes. Presumably, Stephen Colbert understands you can't "cheat death" by eliminating one mortality risk factor. Estimates of increased risk of death from formal scientific survival studies are irrelevant to the likelihood of immortality. They relate to the *relative risk* of dying at a given point in time. We know we will die someday, but we don't want to die now, on this day, this year, or in this TV season of *Ghost Whisperer*. Throughout life, the risk of dying at any point in time depends on age, smoking status, and many other risk factors, including exposure to air pollution.

MORTALITY RISK VERSUS "WE ALL DIE EVENTUALLY"

Teaching university students about mortality risk is interesting. It's common for a student to say, using the false logic of Colbert's comedy, "Don't we all die anyway? Isn't our risk of mortality 100 percent?" Most university students are young and healthy, with an extremely low current risk of dying. Why worry about slight increases in risk when their baseline risk of dying is so low? Many healthy young people think of themselves as invincible. Perceptions of risk change as people get older. Discussing health risk mitigation is perceptibly different with older audiences.

To help illustrate the concept of mortality risk, three survival curves are presented in figure 7.1.[3] The curve with white circles represents a life-table–derived survival curve for a population of never-smokers who live in an area with low air pollution. The curve with gray triangles represents a survival curve for smokers who started smoking at age eighteen. From eighteen years of age, smokers' mortality risk, relative to nonsmokers, is elevated by 100 percent (doubled), which is easily observed. The curve with black circles represents a survival curve applying the same approach to never-smokers in a moderately polluted city; mortality risk is elevated by 20 percent.

It is evident from these curves that age is the dominant mortality risk factor. For never-smokers living in a clean environment, aging increases the risk of dying; few live past 100 years. The median age of death (50 percent survival) can be easily observed. The median age at death for never-smokers is approximately 84.3 years. However, the median age at death is shortened by about six years for smokers and two years for those living in moderate air pollution. Alternatively, at age 84.3 years, when approximately 50 percent of never-smokers are expected to be alive, fewer than 25 percent of smokers would survive.

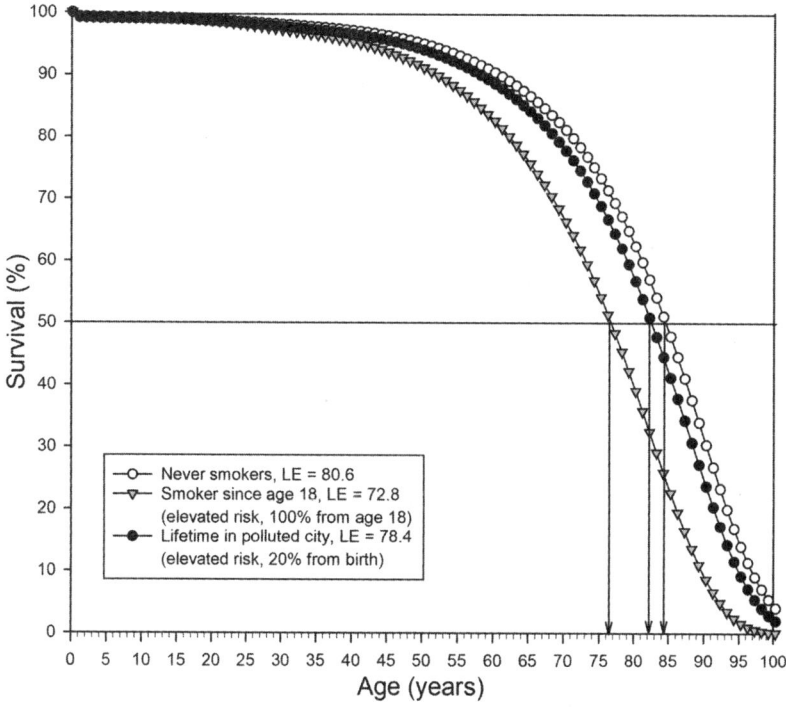

Figure 7.1

Survival curves for never smokers (white circles), smokers since age eighteen (gray tri-angles), and persons living in polluted city (black circles). Drop-down arrows illustrate median age (50 percent survival) at death. *Source:* Modified from C. Arden Pope and Douglas W. Dockery, "Air Pollution and Life Expectancy in China and Beyond," *Proceedings of the National Academy of Sciences of the United States of America (PNAS)* 110, no. 32 (2013): 12861–12862.

One can also integrate data across all ages and calculate average life expectancy. The life expectancy at birth of never-smokers is 80.6 years, while the life expectancy for smokers is 72.8 years, approximately 7.8 years shorter than never-smokers. For never-smokers living in a moderately polluted city with a mortality risk elevated by 20 percent, life expectancy would be 78.4 years, a loss of about 2.2 years attributable to air pollution exposure.

The loss of life expectancy from air pollution is smaller than from smoking. However, smoking affects only a fraction of the population who smoke. You can choose not to smoke and increase your life expectancy. Air pollution exposure, by contrast, is ubiquitous and involuntary. Air pollution affects all breathers of all ages. Because everyone breathes, the overall public health impact of air pollution on attributable mortality may be greater than smoking. As discussed in more detail in chapter 11, the population burden of disease and the resulting loss of life expectancy from air pollution globally is comparable to cigarette smoking.[4]

The bottom line? Exposure to smoking and air pollution both contribute to loss of life expectancy.

AIR POLLUTION INTERVENTION STUDIES FROM IRELAND

As discussed in chapters 4 and 5, multiple studies from the United States, Canada, Europe, and China provided evidence that long-term exposures to $PM_{2.5}$ air pollution are associated with an increased risk of death. Meta-analyses that combine evidence from these studies suggest that a 10 µg/m^3 increase in $PM_{2.5}$ exposure increases death risk by 6–9 percent, reducing life expectancy by approximately a year.[5]

The increased risk of death and shorter life expectancy are primarily due to cardiovascular, respiratory, and lung cancer mortality. This risk increase has substantial public health consequences when applied across large populations, especially for people with relatively high exposures. But there is a catch. These studies also imply that if long-term exposure to $PM_{2.5}$ was reduced, there should be comparable and observable reductions in mortality.

What is the evidence?

Interestingly, the same fundamental question has also been asked about cigarette smoking. Over several decades, evidence mounted that cigarette smoking was a major contributor to death risk—primarily through cardiovascular, respiratory, and lung cancer mortality. Some of the most compelling evidence comes from smoking intervention studies demonstrating that quitting or "smoking cessation" reduces death risk compared with continued smoking. For example, in one major study, quitting smoking substantially reduced the risk of death and resulted in about six to ten years of gained life expectancy compared with continued smoking.[6]

Soon after seeing the results of the cohort studies, Arden and Doug knew it would be helpful to evaluate life expectancy changes when people "quit air pollution" or at least reduce their exposure to air pollution. They understood this would not be as simple as smoking cessation studies, where one can compare smokers who quit to smokers who don't.

In May 1994, Doug was approached at the Annual Conference of the American Thoracic Society in Boston by Luke Clancy, a pulmonologist from Dublin, Ireland. Luke has been a critical player in restricting smoking in workplaces in Ireland and Europe and the ongoing quest for Tobacco Free Ireland. Luke explained that in the 1980s, Ireland had experienced severe air pollution episodes from domestic coal burning, much like the historical London smog episode. He had observed that these episodes exacerbated the respiratory conditions of his patients. Indeed, Luke had published a report that during one particularly severe event in 1982, the fatality rates at a central Dublin hospital doubled.[7] In response, the Irish government banned coal marketing, sale, and distribution in Dublin starting September 1, 1990. Air pollution dramatically improved that winter and the following winter.

This was just the type of situation Doug had been looking for to answer his questions about pollution control and population health. Doug and Luke found that after the ban, particulate matter air pollution was reduced by approximately 36 μg/m³. That represents about a *70 percent reduction*. Adjusted mortality rates dropped by 15.5 percent for respiratory deaths and 10.3 percent for cardiovascular deaths.[8]

The Dublin results were impressive and offered clear evidence of improved health from reduced exposure to particulate matter air pollution. Based on the success of the Dublin ban on coal sales, successive prohibitions were put in place in other Irish cities in 1991, 1992, and 1993, again with evidence of improved air quality.[9]

The coal bans improved respiratory health. However, it was impossible to untangle the role of air pollution from long-term health, social, and economic trends within the study populations, so the researchers could not link reduced overall mortality or cardiovascular mortality specifically to reduced air pollution.[10] Thus, the results from Ireland were suggestive but not conclusive. More evidence was needed.

THE GAMBLER AND A COPPER SMELTER STRIKE

In 1979, John Trijonis published a study with a boring title: "Visibility in the Southwest: An Exploration of the Historical Data Base."[11] John, however, was not boring. He earned a master of science degree in aeronautics in 1967 and a PhD in environmental engineering in 1972, both from Caltech. John was president of the Santa Fe Research Corporation from 1980 to 1993, conducting government-funded studies on air pollution, atmospheric visibility, and related topics. He was also a Las Vegas gambler, specializing in sports betting and mathematical handicapping using large historical databases.

In 2001, the Caltech Alumni Association asked John to describe his life as a Las Vegas gambler. He wrote an article with a less boring title: "Vegas Winners."[12] In that article, he discussed strategies to legally win and earn a good income in the Las Vegas gaming industry. He emphasized the importance of following only legal strategies. He suggested that his scientific training and techniques for analyzing pollution, weather, and visibility data gave him an edge in understanding sports data and competing as a mathematical handicapper. He said that "basically, the winners find a way to be like the casino; they gain a small advantage and play it as often as possible."[13] John explained how his gambling impacted his environmental science career: it provided additional income.

John also shared insights regarding environmental research. He noted that one of the dilemmas he faced as an air pollution scientist was that options for research funding were either government agencies or private industry—the latter often included primary industrial polluters. He believed that "government agencies essentially want to determine if there is a problem and what can be done about it."[14] In contrast, he said:

The polluting industries, on the other hand, often seem in deep denial. They tend to assert that (1) there is no problem, (2) even if a problem exists, they aren't the cause, and (3) even if they were the cause, the nature of the problem is too uncertain to try to do anything about it.[15]

John acknowledged that industry provided the most lucrative consulting contracts. Still, he was a "product of the sixties and a little quixotic"[16] and was most comfortable working on government-funded studies. As such, his income from research projects was limited, and his research results were "subject to incessant carping from industry scientists and their consultants."[17]

John's historical 1979 manuscript on atmospheric visibility in the Southwest described an interesting natural experiment dealing with air pollution. His paper documents a copper smelter strike that occurred across the four southwestern states of New Mexico, Arizona, Utah, and Nevada from July 15, 1967, through the beginning of April 1968. In the 1960s, approximately 90 percent of all sulfate air pollution in these four states came from copper smelters.

The sulfate pollution did not result in extremely high levels of air pollution. Yet it did contribute enough sulfate-based fine particle pollution to pollute the air and influence visibility. Over the eight-and-a-half-month strike period, concentrations of suspended sulfate particles decreased by approximately 60 percent, resulting in slightly less polluted air regionally and improved visibility. John's work did not address health impacts but only evaluated the effect on visibility, particularly in the national parks. But what about health? Did the regional reduction in air pollution also reduce regional mortality? Although this event happened forty years earlier, it was a quasi-experimental research opportunity that needed to be pursued.

Arden had two student research assistants willing to compile historical monthly mortality data by state throughout the United States from 1960 through 1975. Based on John Trijonis' paper, they knew that sulfate particle air pollution was substantially reduced, and visibility improved during the strike. They analyzed the data using a quasi-experimental interrupted time-series study approach. This approach used statistical models controlling for time trends, mortality counts in bordering states, and nationwide mortality for influenza/pneumonia, cardiovascular, and other respiratory deaths. The findings were clear. During the copper smelter strike period, small but measurable decreases in mortality occurred: a 2–4 percent decline. The results were consistent and statistically robust across various models and the four affected states. These results demonstrated that the strike-related reduction in regional sulfate concentrations, which resulted in improved visibility, was also linked to measurable decreases in mortality.[18]

AIR QUALITY STANDARDS: A QUASI-EXPERIMENTAL STUDY MADE IN THE UNITED STATES

In the United States, public policy efforts to reduce air pollution resulted in substantial reductions in overall air pollution in many regions. These efforts included establishing and enforcing the National Ambient Air Quality Standards and specific controls, like the Acid Rain Program, that reduced air pollution from coal-powered power plants.

Reductions in air pollution were not uniform across the United States but have differed across metropolitan areas. These differences were a product of public policy but offered a giant quasi-experiment. From about 1980 to 2000, metropolitan areas, that were initially not in compliance with the air quality standards, reduced their air pollution exposure much more than areas initially in compliance. Therefore, Doug and Arden sought to use this quasi-experiment to test the hypothesis that metropolitan areas with more extensive declines in air pollution would also have larger increases in life expectancy.

This would take a lot of data and work, including controlling for changes in socioeconomic, demographic, and smoking variables. They had access to air pollution data but needed help with the demographic and socioeconomic data. Luckily, they knew Majid Ezzati.

Majid was trained as an electrical and computer engineer before turning his talents to environmental health and risk factors for population health. In 2008, Majid and his research team used individual death records from national mortality statistics and population data from the U.S. Census to calculate annual life expectancies over time for all counties across the United States. He also compiled socioeconomic and demographic data and used these data to evaluate disparities in mortality across time and space.[19]

Doug and Arden, as air pollution researchers, questioned if areas with higher particulate air pollution had shorter life expectancies. They asked Majid if he would share his compiled data and collaborate with them to study air pollution across time and space. Using 211 county units in 51 U.S. metropolitan areas with $PM_{2.5}$ air pollution over two decades, they estimated and published the relationship between air pollution reductions and life expectancy changes.[20]

The results were clear. Counties with larger decreases in air pollution between 1980 and 2000 had greater increases in life expectancy (see figure 7.2). Based on the statistical analysis, a 10 µg/m³ decrease in long-term PM$_{2.5}$ exposures contributed to approximately a 0.6–1.0-year increase in life expectancy compared with counties that had no reduction in exposure.

The bottom line? Reductions in air pollution accounted for substantial improvement in life expectancy over those two decades.

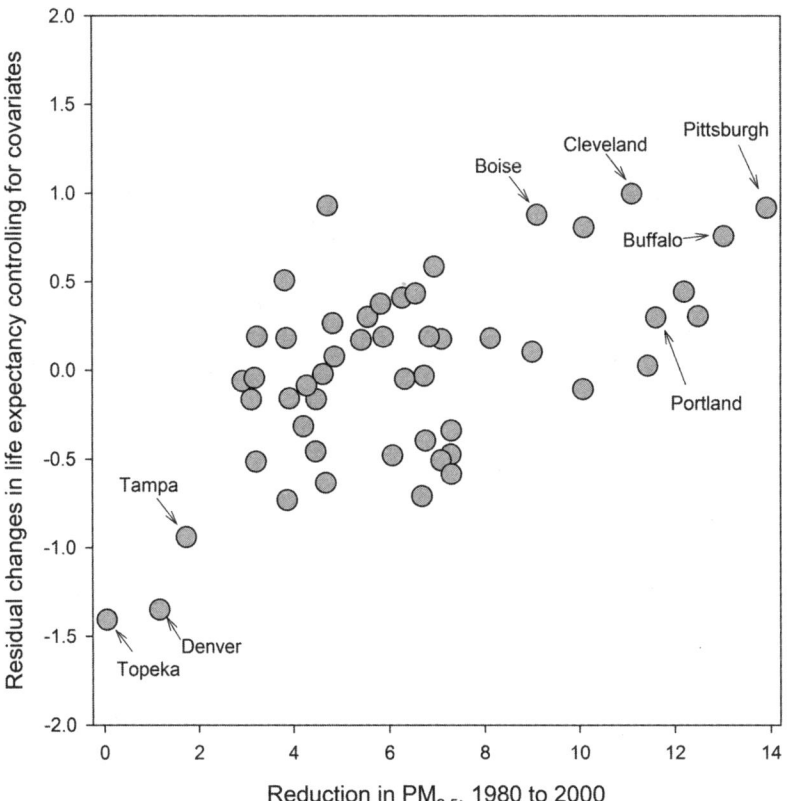

Figure 7.2
Residual changes in life expectancy in the central most populated counties after controlling for changes in socioeconomic, demographic, and smoking variables plotted over reductions in PM$_{2.5}$. *Source:* Illustration created based on data from C. Arden Pope, Majid Ezzati, and Douglas W. Dockery, "Fine-Particulate Air Pollution and Life Expectancy in the United States," *New England Journal of Medicine* 360 (2009): 376–386.

These results were good news. They presented encouraging evidence that improvements in air quality could result in measurable improvements in public health. But not everyone saw it that way. Unsurprisingly, some disputed the results. For example, one contention was that the statistically significant associations between reductions in air pollution and increased life expectancy were due to a single influential observation.[21] This observation was the city of Topeka, Kansas, which had no reduction in pollution, largely because it was a city with low pollution to begin with.

In response, Doug, Majid, and Arden conducted a follow-up study using further statistical analysis. This analysis confirmed the initial results. In fact, the findings of improved life expectancy with reduced air pollution were independent of whether Topeka was included or not, or of any other observation.[22]

Another reanalysis using improved modeled $PM_{2.5}$ exposures found similar results.[23] Francesca Dominici and two of her students extended the analysis to include 545 U.S. counties for 2000–2007. They found that reductions in $PM_{2.5}$, especially sulfates, were associated with increased life expectancy, although the reductions in pollution and the increases in life expectancy were smaller for this shorter period.[24]

Overall, the quasi-experimental analysis of life expectancy in the United States provides evidence that reducing exposure to $PM_{2.5}$ air pollution can substantively improve life expectancy.

A RIVER RUNS THROUGH IT: A QUASI-EXPERIMENTAL STUDY MADE IN CHINA

Michael Greenstone is a natural resource and environmental economist and econometrician. He earned his PhD from Princeton University, was a professor of environmental economics at MIT, and is currently a professor of economics and director of the Energy Policy Institute (EPIC) at the University of Chicago. One of his key contributions to air pollution and other environmental research is an emphasis on the rigorous use of natural experiments and quasi-experimental techniques to support causal inferences more effectively.[25] The 1981–1982 U.S. recession, which resulted in reductions in particulate matter air pollution that varied substantially across sites, provided one opportunity for research. Greenstone and Kenneth Chay used a quasi-experimental approach to estimate that a 1 percent reduction in particulate air pollution was associated with a 0.35 percent decline in infant mortality.[26] This study was interesting, but Greenstone understood the need for additional research, especially in a more polluted area. Therefore, he collaborated on an even more extraordinary quasi-experimental study of air pollution in China.

The Huai River runs from west to east across much of China. When linked with the Qin Mountain Range, it forms the Qinling-Hauihe Line, commonly used to distinguish northern from southern China. In the 1950s, China instituted the Huai River Policy, providing free or highly subsidized coal for indoor heating north of the Qinling-Hauihe Line but not south. Although temperatures are generally colder in northern China versus southern China, there is no sharp discontinuous difference in temperature right at the Huai River. Nevertheless, centralized coal-powered heating systems were developed in cities to the north of the Huai River but not to the south. Significant differences in the use of coal just north and south of the river have persisted over the years.

The Huai River Policy offered a unique quasi-experiment. Was there a rigorous, methodologically sound way to evaluate whether particulate matter air pollution from coal combustion was elevated north versus south of the river? Even more importantly, was there a sharp elevation in mortality or reduction in life expectancy north of the river?

Greenstone and colleagues collected mortality and life expectancy data from the Chinese Center for Disease Control and Prevention. They also collected smoking prevalence, dietary patterns, demographic variables, and other relevant data. They then used a well-developed quasi-experimental econometric analysis approach called regression discontinuity to estimate the difference in air pollution and life expectancy. The elevation in particulate matter air pollution (measured as PM_{10}) just north of the Huai River was approximately 42 µg/m^3. A sharp, discrete decrease in life expectancy of 3.1 years occurs at the Huai River boundary. Additional analysis showed that the reduction in life expectancy was driven by significantly elevated respiratory and cardiovascular mortality, not deaths from other causes.[27]

The striking results of the Huai River study were highly robust. They implied massive adverse health effects of exposure to air pollution, especially in a country with high pollution levels. But, in per-unit exposure to air pollution, were these results comparable to those observed in the less-polluted United States?

The Huai River study indicated that a 10 μg/m³ increase in long-term exposure to PM_{10} reduced life expectancy by 0.64 years. Assuming that approximately half of PM_{10} is $PM_{2.5,}$ these results suggest that a 10 μg/m³ increase in long-term $PM_{2.5}$ exposure would reduce life expectancy by about 1.28 years. These estimated effects are similar and, in fact, a bit larger than those observed from the quasi-experimental study in the United States discussed above.

Michael Greenstone and his EPIC colleagues used the results of the Huai River study to create an Air Quality Life Index. Estimates based on this index indicated that particulate matter air pollution was a considerable threat to human health and one of the largest contributors to loss of life expectancy worldwide.[28]

SOUTHERN CALIFORNIA CHILDREN'S HEALTH STUDY

As we've already discussed, there is substantial evidence that reducing exposure to air pollution reduces mortality risk and increases life expectancy. Is there also evidence that reduced air pollution improves health and quality of life? What are the effects of air quality improvements on children's health?

Among the most important and impressive ongoing efforts to study the health effects of air pollution is the University of Southern California Children's Health Study. This ambitious research effort was initiated in 1996 by John Peters, who had a distinctive background. After earning a medical degree from the University of Utah, and doing surgical residency at Johns Hopkins University, he earned a doctor of science at Harvard studying with Ben Ferris. He was on the faculty of the Harvard School of Public Health before moving to the University of Southern California.

The Southern California Children's Health study was designed to study the effects of chronic or long-term exposure to air pollution on the developing lungs of schoolchildren. As such the initial and ongoing research has been conducted over the years by various remarkable collaborators, including James Gauderman, Edward Avol, Kiros Berhane, Frank Gilliland, Duncan Thomas, Rob McConnell, Erika Garcia, and others.

The study enrolled thousands of children in kindergarten and first grade and followed them into young adulthood. Over the years, the study reported many essential findings. For example, children living in areas with higher regional air pollution had significant deficits in lung development from age ten to eighteen.[29] The same was true for children with more exposure to traffic-related air pollution from freeways.[30]

The Southern California Children's Health study also made important findings directly related to our central question: Does reducing air pollution improve health? It found that reductions in air pollution were linked with several measures of improved health in children:

1. Air pollution reductions "were associated with statistically and clinically significant positive effects on lung function growth in children."[31] Doug and Jim Ware wrote an editorial on the importance of and strengths of this research, aptly titled "Cleaner Air, Bigger Lungs."[32]
2. Reductions in air pollution were associated with fewer bronchitis cases and related respiratory symptoms in children.[33]
3. Reductions in air pollution were associated with fewer asthma cases in children.[34]

The results of the Southern California Children's Health study are clear. Exposure to air pollution has measurable and significant adverse impacts on children's respiratory health. But, happily, reduced exposure to air pollution improves children's respiratory health.

WOULD IT HELP TO MOVE?

Another way to reduce air pollution exposure is to move to a cleaner location. But is there evidence that moving reduces the adverse effects of air pollution?

One revealing study included children who were part of the Southern California Children's Health Study. Children who moved from Southern California to areas with different levels of air pollution were followed up. Based on critical measures of lung function, as a group, children who moved to areas with lower air pollution experienced more significant growth in lung function. However, children who moved to areas with higher air pollution experienced relatively slower growth in lung function. The differences in lung function growth were most pronounced for those who had relocated at least three years before the follow-up visit.[35] These results demonstrate that relocating to a place with less-polluted air benefits children's lung health.

Another interesting quasi-experimental study evaluated changes in exposure to $PM_{2.5}$ pollution after moving. This study used data from some of Canada's largest and most well-compiled cohorts, as discussed in chapter 5. The Canadian research team gleaned data from cohorts of adults who had consistently lived in high- or low-pollution areas and then moved during the ensuing five years. The study included nearly 700,000 adults.

The researchers compared their mortality rates from various causes with changes in long-term exposure to $PM_{2.5}$ air pollution. Five years after moving, those who moved to a less-polluted area experienced a significant reduction in mortality. Greater declines in mortality were observed among those with the most pollution reduction. Increased mortality occurred among those who moved from low- to high-pollution areas. The authors concluded: "In Canada, decreases in $PM_{2.5}$ were associated with lower mortality, whereas increases in $PM_{2.5}$ were associated with higher mortality. These results were observed at $PM_{2.5}$ levels considerably lower than many other countries, providing support for continuously improving air quality."[36] The title of an accompanying editorial by environmental health researcher Gavin Pereira sums up the results: "Cut Particulate Air Pollution, Save Lives."[37]

CONCLUSION

"Does reducing air pollution improve health and reduce mortality?"

Yes, it does, whether pollution sources are removed or controlled, or people move away from the pollution. While it is difficult to disentangle the effects of reduced air pollution from other factors that affect human health and mortality, the overall evidence is quite compelling.

The studies from Meuse Valley, Donora, and London discussed in chapter 2 provide evidence of the harmful effects of air pollution. During the dramatic air pollution episodes addressed in chapter 2, there was an apparent increase in disease and death. Still, the elevated disease and death rates returned to lower levels when the air pollution episodes subsided. The Geneva Steel natural experiment studies covered in chapter 3 showed that health improved with reduced air pollution exposure during the mill's temporary closure. The time-series and case-crossover studies discussed in chapter 3 linked increased disease and death with increased short-term pollution exposure.

All these studies observed decreased illness and death with subsequent reductions in exposure. Even the cohort studies of long-term exposure and risk of mortality discussed in chapters 4 and 5 provided evidence. For example, as discussed in chapter 4, an extended follow-up analysis of the Harvard Six-Cities study found that lower air pollution levels were associated with lower mortality risk over time, providing evidence that $PM_{2.5}$-associated mortality was partially reversible.[38]

The studies discussed in this chapter provide some of the most powerful evidence. These natural, quasi-experimental studies offered valuable research opportunities. They include natural experimental studies of coal bans in Ireland and a copper smelter strike in the U.S. Southwest, a quasi-experimental study of air pollution reductions due to policy implementation, a unique regression discontinuity study of the Huai River policy in China, a well-designed and implemented longitudinal study of children's health in Southern California, and straightforward relocation studies of lung development in U.S. children and mortality in Canadian adults.

As John Trijonis might say, these studies have been subject to "carping from industry scientists and their consultants."[39] Nevertheless, these results point to a fundamental conclusion: reducing exposure to $PM_{2.5}$ air pollution improves public health.

8

IF AIR POLLUTION IS SO LETHAL, WHY ISN'T EVERYONE DEAD?

Air pollution research should be scrutinized, questioned, validated, and explored because this research has public policy, public health, and even economic implications. There is clear value in verifying and understanding this research and its implications. However, one of the challenges of conducting such research over the last several decades has been addressing legitimate scientific uncertainty while avoiding distracting criticism—the kind created specifically to manufacture doubt. Sometimes it is difficult to determine which is which.

For example, critics of air pollution and health research have emphasized the extreme differences between $PM_{2.5}$ exposure from active smoking compared with $PM_{2.5}$ exposure from air pollution. This contrast is illustrated in figure 8.1. A large bag of sugar, representing the cumulative $PM_{2.5}$ exposure of a smoker, sits next to a couple of small sugar pouches, representing the cumulative lifetime exposure to ambient $PM_{2.5}$ from ambient air pollution.[1]

The essential argument is that, given the minimal exposure to $PM_{2.5}$ air pollution compared with smoking, health effects from air pollution should be nonexistent or extremely small. Or, if the impact of $PM_{2.5}$ is so significant, given the massive $PM_{2.5}$ exposures from active cigarette smoking, why isn't everyone who smokes dead? Is this a valid concern that

Figure 8.1
Figure used by critics to illustrate differences in long-term PM$_{2.5}$ exposure from active smoking versus air pollution. *Source:* Adapted from a color image from comments submitted by the Competitive Enterprise Institute (CEI) regarding EPA's proposed rule entitled, "Reconsideration of the National Ambient Air Quality Standards for Particulate Matter" (Docket ID No. EPA-HQ–OAR-2015-0072), posted by the EPA on April 4, 2023, https://www.regulations.gov/comment/EPA-HQ-OAR-2015-0072-2483.

should be addressed, or is it just a clever attempt to confuse the matter, obstructing the reduction and control of air pollution?

SCIENTIFIC RIGOR OR DISTRACTING CRITICISM?

Doug and Arden have often tried to address scientific uncertainty while facing distracting criticism. As an Interdisciplinary Program in Health fellow at the Harvard School of Public Health, Arden was set up in an excellent research space on the thirteenth floor with a computer terminal connected to the mainframe computer. (This was before laptops.) His office looked out over Boston and Harvard Medical School. Doug's office was just down the hall. Arden remembers well a morning in 1993 when

Doug unexpectedly entered his research room. Doug, who is usually calm, was ticked off. He had just read some harsh criticism that misrepresented their research. This was nothing new. Over the years of doing air pollution and health research, they became increasingly well-acquainted with criticism and misrepresentation of their research.

But what should they do about it? Should they react with a sharp response? They rarely made sharp responses. They concluded that their primary response would be ongoing efforts to conduct the best research possible.

Knowing which scientific and public health controversies involve legitimate scientific inquiry and which are based on ulterior motives can be challenging. On the surface, they may look the same. We must look deeper—to motivation, especially—to discern the difference.

It is expected that scientists will disagree about the conduct, interpretation, and application of research. This is an essential part of science. Addressing these differences is essential and ultimately strengthens science. For example, there were legitimate controversies about how best to address weather variables in the daily time-series mortality studies, how to best design case-crossover analyses, or how to statistically adjust for age and cigarette smoking in the cohort studies. Calls to replicate, validate, and even fine-tune studies are legitimate and should be addressed.

However, effective and efficient science also requires researchers to avoid distractions created by controversy and criticism that try to "muddy the waters" with manufactured doubt. When published studies have been independently reanalyzed and reproduced, and dozens of other studies have found and published similar results, what is the motivation to brand these studies as "secret science"? Why try to prohibit their consideration in the public health policy debate because the researchers protect the privacy and confidentiality of research participants?

Some efforts to manufacture doubt include personal attacks and accusations directed at individual air pollution researchers. One egregious example includes allegations of scientific misconduct directed at Herbert Needleman, a pediatrician, child psychiatrist, and medical researcher at Boston Children's Hospital, Harvard Medical School, and the University of Pittsburgh School of Medicine. Why was he accused of scientific misconduct (including inadequate data analysis and misrepresentation of results)? In short, he conducted groundbreaking research that found that even relatively low levels of lead exposure were enough to adversely affect children's cognitive abilities—and these results threatened the lead industry.[2] Needleman was eventually cleared of the scientific misconduct charges. Subsequent analyses of his data, including those conducted by Joel Schwartz, found the results robust. Further independent research has confirmed Needleman's initial findings.[3]

Various air pollution and health researchers have been accused of scientific or research misconduct. These included researchers who conducted the early CHESS studies,[4] the Harvard Six-Cities and ACS CPS-II cohort studies,[5] and the U.S. Medicare cohort studies.[6] A year-long National Institutes of Health (NIH) investigation regarding the Harvard Six-Cities study found no basis for misconduct allegations.[7] To the authors' knowledge, no university investigation, Office of Research Integrity, or other formal inquiry has found any merit to these accusations.

Carolyn Kormann wrote an article for the *New Yorker* that included interviews with Doug and Steve Milloy. Steve Milloy is a lawyer and lobbyist who served as a member of the EPA transition team for the Donald Trump presidential administration. He is also the founder and publisher of JunkScience.com, a website that tries to dismiss as "junk science" much environmental science related to air pollution, secondhand cigarette smoke, ozone depletion, and climate change.

The article documented Doug's leadership role in the Harvard Six-Cities study and discussed the EPA's proposed "Strengthening Transparency in Regulatory Science," or "secret science" rule, and related public policy issues. Kormann also documented Milloy's ties to the tobacco, coal, and fossil fuel industries and quoted him as saying, "I'm all for the coal industry, the fossil fuel industry. Wealth is what makes people happy, not pristine air, which you'll never get."[8]

Some criticisms of the air pollution and health research seemed to be superficial and designed to simply be distracting. For example, in 1997, Milloy published a brief article titled "Pope-a-Dope?"[9] This article referenced Muhammad Ali's 1974 heavyweight boxing match with George Foreman. Ali used a "rope-a-dope" strategy, leaning back against the ropes, allowing Foreman to tire himself out with his punches. Arden, Doug, and research colleagues had recently published the ACS CPS-II cohort study,[10] and they were careful to protect confidential research data. They were also cautious about extrapolating the results from this study to estimate deaths worldwide.[11] They preferred approaching scientific uncertainties and disagreements by patiently and carefully conducting research and publishing in the peer-reviewed literature. They tried to avoid academic "slug fests."

Milloy, however, derisively implied that their approach was akin to Ali's rope-a-dope strategy. How should they respond to this absurd assertion? Muhammad Ali was one the greatest boxers of all time; they were a couple of university professors and persistent academic researchers. Although absurd, if someone wanted to compare them with one of the world's greatest boxers, fine.

A LEGITIMATE CONTROVERSY

What about the large bag of sugar versus a couple of small sugar packets introduced at the beginning of this chapter and illustrated in figure 8.1? Was this another absurd comparison, or did it illustrate a valid concern that should be seriously addressed? In about 2007, Steve Packham, a toxicologist at the Utah Division of Air Quality and later a member of the U.S. EPA Clean Air Scientific Advisory Committee, gave a presentation outlining the issue of massive differences in exposures to $PM_{2.5}$ from active smoking versus ambient air pollution. Although Doug and Arden had been aware of criticism on this issue for several years, Steve's presentation piqued a desire to address and try to understand this issue and treat it as a legitimate concern. What could be learned by exploring this question?

Figure 8.2 helps illustrate this intriguing issue.[12] Estimates of the adjusted relative risk of cardiopulmonary mortality for pack-a-day smokers (relative to never-smokers) are plotted against estimates of the daily dose of $PM_{2.5}$, the cumulative mass of particles delivered to the lung daily. The relative risk of smoking is approximately 2.0, with a massive $PM_{2.5}$ dose of roughly 240 mg. Estimates of the adjusted relative risk of cardiopulmonary mortality from air pollution exposure in a polluted city or secondhand cigarette exposure from living with a smoking spouse are plotted against estimates of the daily dose of $PM_{2.5}$. The relative risk from living in a polluted city or living with a smoking spouse is much smaller, as is $PM_{2.5}$, compared with smoking.

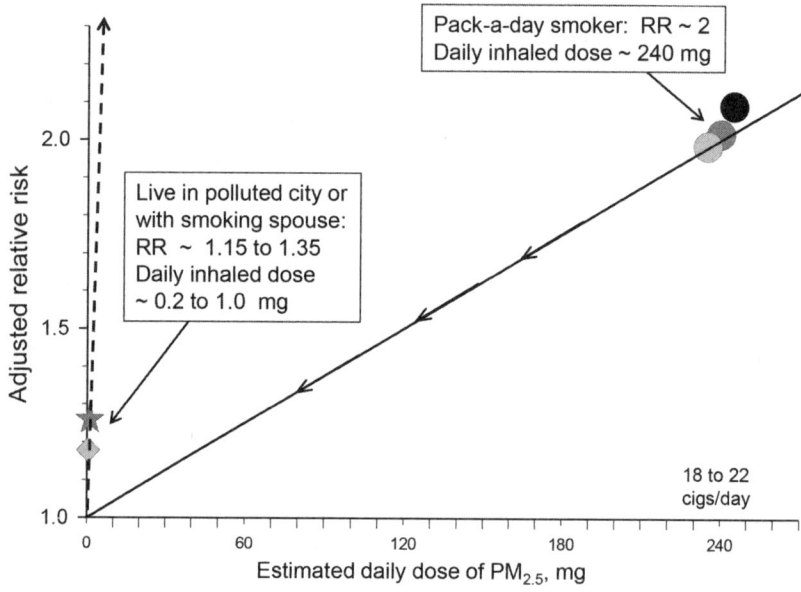

Figure 8.2

Stylized illustration of the dilemma regarding risks from active smoking versus air pollution or secondhand smoke.

So, what is the dilemma? We can see from the solid black line in figure 8.2, assuming a *linear exposure-response function* and extrapolating down from the smoking estimates, that the effects of air pollution (and secondhand cigarette smoke) should be nearly zero. But they are much larger than expected, given their relatively low dose. Or, as shown by the dashed black line, we can alternatively assume a linear relationship and extrapolate up from the estimates of secondhand smoke or air pollution. When these estimates are extrapolated up to the massively larger dose from smoking, nearly all who smoke should be dead.

What was going on?

INTEGRATING EVIDENCE FROM AIR POLLUTION AND CIGARETTE SMOKE

Working with Rick Burnett, Arden explored the dilemma using actual data. It's one thing to make assumptions regarding the shape of the

PM$_{2.5}$-mortality dose-response relationship, but another to estimate it with real data. In collaboration with Michael Thun and other ACS researchers, Arden, Rick, and colleagues used data from the ACS CPS-II cohort. They analyzed the shape of the dose-response between cardiopulmonary mortality and cigarette smoking.[13]

The empirical analyses—looking only at different levels of cigarette smoking—demonstrated that the dose-response function for cardiopulmonary deaths was *not* a linear relationship that started at the origin. The empirical evidence for cigarette smoking demonstrated that *even at the lowest levels of cigarette smoking*, the mortality risk was highly elevated. A linear fit through the active smoking data would start not at the origin but at a high elevated risk. Furthermore, mortality risk was elevated even for the much smaller exposures from secondhand cigarette smoke.

It was as if there were some striking, discrete, harmful consequence of going from a nonsmoker to very light smoking. Did this jump in risk start at just one puff a day? Did increased risk occur with secondhand smoke? Elevated risks from secondhand smoking were observed at extremely low exposures compared with active smoking.

The team conducted additional analyses exploring the shape of the PM$_{2.5}$-mortality exposure-response relationship by integrating evidence from exposures to smoking, secondhand smoke, and air pollution.[14] They generated risk functions to estimate the global disease burden across a wide range of cumulative exposures.[15] The findings were remarkable. The slope of the exposure-response function was very steep for exposures from secondhand smoke, air pollution, and even very light smoking, but leveled off at exposures from moderate to heavy smoking.

Figure 8.3 illustrates the relationship between exposure to PM$_{2.5}$ and cardiovascular and cardiopulmonary disease mortality that integrates data from smoking, secondhand smoke, and air pollution. Exposure and risk estimates for active smoking, secondhand smoke, and air pollution, based on multiple studies, are plotted together.[16] Panel A presents the results on a linear scale, and panel B shows the results on a log scale.

Why plot this mortality exposure-response relationship over two scales? Because the linear relationship implies that the increase in mortality risk is approximately constant for all incremental increases in exposure throughout the full range of exposure. This implication may not be

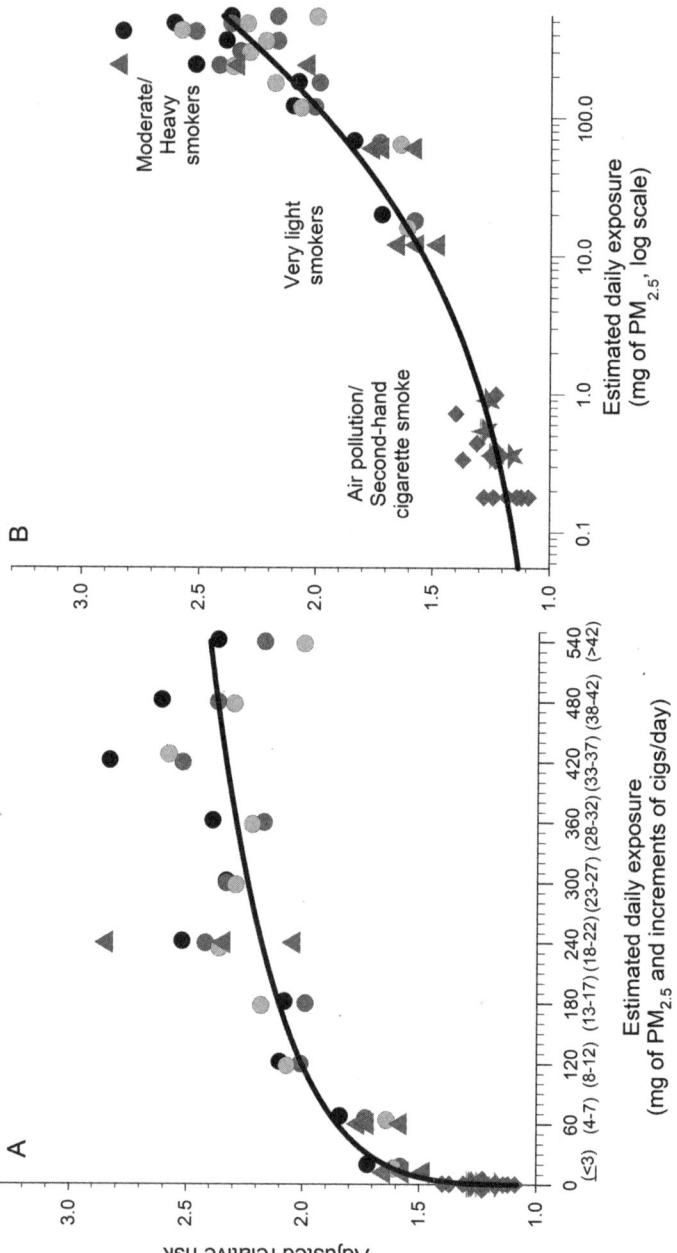

Figure 8.3

Illustration of the integrated exposure-response function for PM$_{2.5}$ and cardiopulmonary or cardiovascular mortality using estimates for exposure to active smoking (black and gray circles and triangles), secondhand smoke (gray stars), and air pollution (gray diamonds) plotted on a linear scale (panel A) and on the log scale (panel B). *Source:* Modified from C. Arden Pope, Aaron J. Cohen, and Richard T. Burnett, "Cardiovascular Disease and Fine Particulate Matter," *Circulation Research* 122, no. 12 (2018): 1645–1647.

suitable for such a large range of exposures. An alternative approach is to plot the increase in risk of $PM_{2.5}$ on the log scale, which implies that relative or percentage increases in exposure are most relevant.

The overall findings illustrated in figure 8.3 are revealing. The $PM_{2.5}$-mortality exposure-response relationship (as shown in panel A) does not appear to be linear across the range of exposures from different sources. Instead, the exposure-response curve is steep at low exposures and then levels off at higher exposures. The $PM_{2.5}$-mortality exposure-response relationship (as shown in panel B) closely, but not perfectly, fits a log-linear response relationship. These findings suggest that a linear exposure-response relationship is not appropriate over this extremely large range of exposure and that relative or percentage increases in exposure are most relevant to increased risk.

These results are not necessarily surprising. Exposure-response relationships in biological processes often better fit a log-linear relationship, especially of large ranges of exposure. The biological processes related to air pollution and health will be discussed briefly below but in more detail in chapter 10. However, the results illustrated in figure 8.3 demonstrate that the overall $PM_{2.5}$-mortality relationship to air pollution, secondhand smoke, and cigarette smoking seems to be exposure-dependent, consistent, and coherent—but not on a linear dose-response scale. These results also illustrate that there is *no safe threshold* for cigarette smoking and that the risk from secondhand exposure is similar to that observed with air pollution.

GLOBAL EXPOSURE MORTALITY MODELS

Exploring the $PM_{2.5}$-mortality exposure-response relationship had more relevance than just addressing critics; it also had an important practical use. The estimated relationship could be used to extrapolate health impacts more reasonably across different levels of exposure, especially for populations with very high pollution exposure levels.

The team's studies evaluated the shape of the $PM_{2.5}$-mortality exposure-response relationship by integrating estimates from both air pollution and cigarette smoke. These studies did not assume that the exposure-response relationship was linear.[17] However, this integrated approach created another potentially problematic assumption. It assumed the same toxicity of PM$_{2.5}$ from different sources. The integrated approach does not allow for potential differences in toxicity that may depend on the exposure source.

For example, PM$_{2.5}$ air pollution from burning coal, gasoline, diesel, and other fossil fuels that are primary contributors to air pollution may be more toxic than PM$_{2.5}$ from tobacco smoking. Similar exposures and PM$_{2.5}$-mortality associations occur for PM$_{2.5}$ from air pollution and secondhand smoking—suggesting similar toxicities.

In response, Rick Burnett led two research efforts that focused on estimating the exposure-response relationship using only exposure from ambient air pollution. Rick established collaborations with fifteen research groups worldwide that had conducted cohort studies on long-term exposure to outdoor PM$_{2.5}$ and mortality, including crucial studies

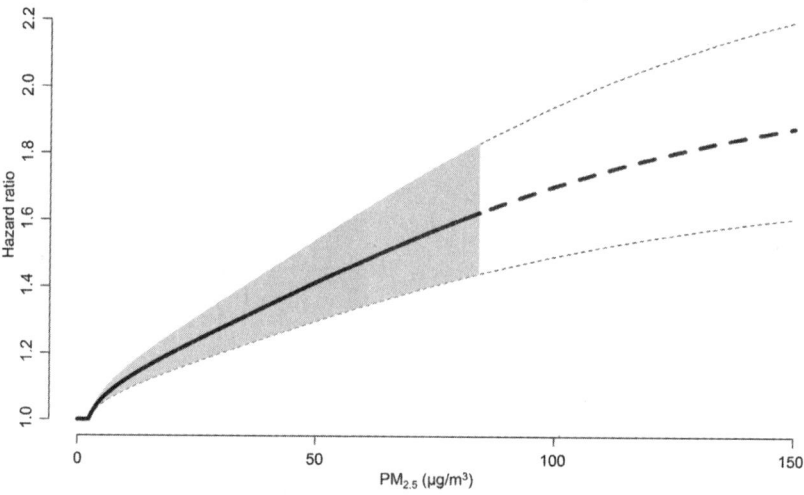

Figure 8.4

Illustration of the exposure-response relationship between PM$_{2.5}$ air pollution and mortality using pooled data from forty-one air pollution studies across sixteen countries. Predictions over the observed concentration range (solid line), extrapolation beyond the range of exposure (dashed line), and 95 percent uncertainty interval (dotted lines). *Source:* Adapted from Richard T. Burnett, Hong Chen, Mieczysław Szyszkowicz, Neal Fann, Bryan Hubbell, C. Arden Pope, Joshua S. Apte, et al., "Global Estimates of Mortality Associated with Long-Term Exposure to Outdoor Fine Particulate Matter," *Proceedings of the National Academy of Sciences* 115, no. 38 (2018): 9592–9597 (figure S9).

discussed in previous chapters. They collected results from an additional twenty-six cohort studies. Overall, the team had data from forty-one different cohorts from sixteen countries. They pooled information from all these studies and developed a "global exposure mortality model" for various causes of death.[18]

The estimated exposure-response relationship for all-cause mortality is presented in figure 8.4. The results demonstrate that, within the range of exposures that are observed in air pollution studies, a near-linear relationship is observed, but with evidence of a leveling off at higher exposures.

An additional approach compiled $PM_{2.5}$-mortality relationships reported from cohort studies in numerous countries and used a sophisticated, flexible risk function in a meta-analytic modeling framework.[19] The estimated $PM_{2.5}$-mortality exposure-response relationships for outdoor air pollution were similar for both research efforts that used only exposure from ambient air pollution. They were *not* linear over the full range of exposures; instead, they were also relatively steep at lower exposures and leveled off at higher exposures.

PRODUCTIVE RESPONSE TO RELEVANT CONTROVERSY

In this chapter, we addressed controversial efforts to invalidate or discount evidence that $PM_{2.5}$ air pollution contributes to the risk of disease and death. One argument is that because $PM_{2.5}$ exposure from air pollution is so minimal compared with active smoking, health effects from air pollution should be nonexistent or extremely small. But if the health effects from air pollution and secondhand smoke were real, you might also ask, why isn't everyone who smokes dead?

It turns out that this controversy is based on a fabricated dilemma that occurs if you assert two underlying assumptions: (1) There is equal toxicity of $PM_{2.5}$ exposures from air pollution and cigarette smoking, and (2) the $PM_{2.5}$-mortality association is strictly linear across the full range of exposure from ambient air pollution, secondhand smoking, and active smoking. If either or both assumptions are false, there is no dilemma.

Cigarette smoke consists of many harmful and potentially harmful components.[20] Although the contributions of specific components are not fully understood, it is well established that exposure to cigarette smoke contributes to the risk of respiratory, cardiovascular, lung cancer, and other diseases. $PM_{2.5}$ air pollution also consists of complex mixtures of harmful components emitted from diesel and gasoline vehicle tailpipes, smokestacks of coal-fired power plants and high-temperature industrial processes, and other sources.

Using comparable exposure units, the toxicity of $PM_{2.5}$ in air pollution may be even more significant than $PM_{2.5}$ in tobacco smoke. Comparing the estimated effects of long-term exposure to air pollution with those of secondhand tobacco smoke suggests similar toxicity (see figure 8.3). $PM_{2.5}$ from air pollution and secondhand smoke may thus be more toxic on a per-unit basis than active smoking.

Further, there is no reason to assume strict linearity across such a wide range of exposures. The empirical evidence is consistent with risk functions that are steeper at lower exposures and flatten out at higher exposure. This finding suggests that relatively low $PM_{2.5}$ exposures can activate or initiate relevant biological pathways, such as pulmonary and systemic inflammatory responses. These biological pathways will be further discussed in chapter 9.

Increased exposure raises risks but at a decreasing marginal rate. For example, exposures to secondhand smoke or ambient air pollution appear sufficient to initiate adverse biological responses. Smoking just a cigarette or two daily provides ample exposure to trigger adverse biological responses. However, effects increase more slowly at higher rates of exposure. That is why going from zero to three cigarettes per day results in an incremental risk increase far more significant than going from twenty to twenty-three cigarettes per day.

CONCLUSION

"If air pollution is so lethal, why isn't everyone dead?"

Because the exposure-response relationship is not linear. The incremental health effects of air pollution—smoking included—level off at higher exposures. It's established that cigarette smoking exposes the body to high doses of harmful substances, substantially contributing to an elevated risk of disease and death. Even nonsmokers exposed to secondhand smoke are subjected to adverse health risks. And just like secondhand smoke, ubiquitous and involuntary exposures to air pollution also impose substantive negative public health impacts.

Efforts to directly address the controversy regarding the dramatic differences in exposure to $PM_{2.5}$ from air pollution versus active smoking provided important insights into the health effects of particle pollution. They helped provide better estimates of the shape of the relationship between mortality and air pollution. These efforts have also been extremely helpful in making more informed estimates of the contributions of air pollution to global disease and deaths.

9

DOES EPIDEMIOLOGICAL RESEARCH ON AIR POLLUTION PROVIDE EVIDENCE OF A CAUSAL RELATIONSHIP?

During the Trump administration, the EPA's process of reviewing national air quality standards was thrown into chaos. Industry-funded consultant Louis Anthony "Tony" Cox was appointed chair of the Clean Air Scientific Advisory Committee (CASAC), a key EPA advisory committee that provides consultation and peer review on air quality standards. Cox earned his PhD in risk analysis from MIT. He is president of Cox Associates Consulting in Denver, Colorado. Cox Associates provides consulting services to various industry groups, including the American Petroleum Institute, the American Chemistry Council, the National Mining Association, and Philip Morris International. When determining the health effects of air pollution, Cox emphasizes uncertainties inherent in scientific analysis and casts doubts on the evidence. Cox's critics argue that much of the uncertainty he produces is manufactured and influenced by industrial polluters.[1]

Tony Cox's appointment to chair CASAC generated concern that scientific-based environmental and public health policy was "under siege."[2] This concern stemmed partly from Cox's arguments that air pollution research studies dealing with the health effects of $PM_{2.5}$ should be discarded for use in policymaking or in setting air quality standards. This concern grew when the newly appointed EPA administrator dismissed qualified scientific leaders from CASAC.

Cox also dismissed the panel of twenty academic and industry scientists appointed to review the scientific evidence on national standards for $PM_{2.5}$. One of the academic scientists dismissed was Doug.

CORRELATION IS NOT CAUSATION

Foundational to Cox's criticisms of air pollution and health research was an argument that the epidemiologically observed associations between air pollution and disease and death did not prove causation. The argument that *correlation is not causation* is sometimes used to simplistically dismiss or marginalize evidence that air pollution contributes to respiratory and cardiovascular disease and death.[3]

So, what are the most important things to know about the assertion that "correlation is not causation"? First and foremost, *it is true*. Correlations or associations alone do not prove causation. The imaginary research project that observed correlations between ice cream sales and shark attacks is a classic illustration of this principle. Naive researchers might erroneously interpret these correlations as causal. They might even recommend avoiding ice cream while at the beach.

Thoughtful researchers would be cautious about making causal inferences. They would find the results interesting but would conduct additional research. They might find that ice cream sales were associated with and even caused by warm weather. With more studies, they might find that warm weather was correlated with ice cream sales and *causally* associated with swimming in the ocean while sharks were active near the shore. In other words, shark attacks occur more often when the weather is warm because when the temperature is higher more people are cooling off in the ocean. They're also more likely to be enjoying ice cream.

The second thing to know about "correlation is not causation" is that this issue has long been well-known, discussed, and debated among medical researchers, epidemiologists, biostatisticians, econometricians, and others.[4] Appreciation of this issue is a primary driving motivator in the design, implementation, and interpretation of observational research, including research on the health effects of air pollution.

MANIPULATIVE CAUSATION: A CONTROVERSIAL FIGURE AT THE EPA

In the case of Tony Cox, the controversial chair of CASAC from 2017 to 2021, his arguments about correlation and causation were couched in relatively thoughtful and sophisticated analyses.[5] There are many examples in the literature where associational and causal interpretations *should* be more rigorously scrutinized. What detracted from Cox's credibility while on CASAC was that he largely ignored existing concerted efforts by many skilled air pollution researchers, including statisticians and econometricians, to address issues regarding causal inference. Cox promoted narrow, limited, and restrictive methodological approaches he described as "manipulative causality" that could specifically be applied using the "Causal Analytic Toolkit (CAT)" developed and endorsed by Cox Associates Consulting.[6] He often criticized alternative statistical and econometric methods.

Cox's focus on manipulative causation contrasted with efforts that examined the larger body of research in broader, more inclusive "weight-of-the-evidence" or *systematic-review* approaches. While he served as chair of CASAC, some researchers were concerned that his approach was less about a comprehensive evaluation of evidence and more about weakening the process of setting and reviewing air quality standards. For example, Gretchen Goldman and Francesca Dominici argued that we should be careful about efforts to establish causality and that we should not abandon evidence and process with regard to air pollution policy.[7]

Early in the Biden administration, Cox was replaced as the chair of CASAC. Scientific inferences about causality using broader weight-of-the-evidence or systematic-review approaches resurfaced as valid ways to explore the evidence. Weight-of-evidence and systematic-review approaches are increasingly informed by study designs and methods that enhance the ability to make causal inferences.

It might be tempting to suggest that growing interest in causal inference methods resulted from Cox's high-profile emphasis on manipulative causation. Probably not so much. The growing use of rigorous causal methods and efforts to evaluate the health benefits of reduced air pollution began well before Cox was appointed chair of CASAC.

Ironically, the air pollution researchers most critical of Cox's focus on manipulative causation had long advocated for and used a wide array of study designs that enhanced the ability to make causal inferences. These researchers respect and understand the value of well-designed and well-conducted panel studies, time-series studies, prospective cohort studies, and other epidemiological studies. They also recognize and seek opportunities to conduct studies that provide more robust causal inferences. Doug and Arden are among these researchers, as are Francesca Dominici, Corwin Zigler, Joel Schwartz, and many others.[8]

As scientists and researchers, we understand that correlations are not conclusive proof of causation. It is essential, however, to understand that this does not negate the value of well-designed and well-conducted analyses of observational data—that is, rigorous epidemiology. Epidemiologic evidence based on real people experiencing actual exposures can be highly informative and useful in identifying and quantifying the risk of environmental contaminants. The question is, How do you evaluate the strength of the evidence from epidemiologic studies?

There is no single magic solution to establishing causality using real-world observational data. In the real world, causal relationships are often complicated, with multiple causal and non-causal factors and characteristics that may correlate with each other, resulting in potential confounding, just as they did in the imaginary case of the shark bites and ice cream.

BRADFORD HILL CRITERIA FOR CAUSATION

Austin Bradford Hill was a pioneering epidemiologist and professor of medical statistics. Working with Richard Doll, Hill was instrumental in discovering the link between smoking and lung cancer in the early 1950s. In 1965, he presented an influential address titled "The Environment and Disease: Association or Causation," published in the *Proceedings of the Royal Society of Medicine*.[9] Hill proposed a set of nine criteria to provide epidemiologic evidence of a causal relationship between a presumed cause and an observed effect. These Hill criteria are outlined in box 9.1.

Box 9.1
Bradford Hill criteria for causation in epidemiologic studies

- Strength (effect size): A small association does not mean that there is not a causal effect, though the larger the association, the more likely that it is causal.
- Consistency (reproducibility): Consistent findings observed by different persons in different places with different samples strengthen the likelihood of an effect.
- Specificity: Causation is likely if there is a very specific population at a specific site and disease with no other likely explanation. The more specific an association between a factor and an effect is, the bigger the probability of a causal relationship.
- Temporality: The effect has to occur after the cause (and if there is an expected delay between the cause and expected effect, then the effect must occur after that delay).
- Biological gradient (dose-response relationship): Greater exposure should generally lead to a greater incidence of the effect. However, in some cases, the mere presence of the factor can trigger the effect. In other cases, an inverse proportion is observed: greater exposure leads to lower incidence.
- Plausibility: A plausible mechanism between cause and effect is helpful [but Hill noted that knowledge of the mechanism is limited by current knowledge].
- Coherence: Coherence between epidemiological and laboratory findings increases the likelihood of an effect. However, "lack of such [laboratory] evidence cannot nullify the epidemiological effect on associations."
- Experiment: "Occasionally, it is possible to appeal to experimental, or semi-experimental evidence."
- Analogy: The use of analogies or similarities between the observed association and any other associations.

Although there have been substantial advances in evaluating causal associations, the Hill criteria remain foundational for assessing causation in epidemiologic studies nearly six decades later. Interestingly, Hill, a statistician, did not include statistical significance among these criteria. Indeed, he wrote:

No formal tests of significance can answer those questions. Such tests can, and should, remind us of the effects that the play of chance can create, and they will instruct us in the likely magnitude of those effects. Beyond that, they contribute nothing to the "proof" of our hypothesis.[10]

WEIGHT-OF-THE-EVIDENCE APPROACH

In the United States, the Environmental Protection Agency (EPA) is charged with identifying air pollutants that could affect public health and establishing National Ambient Air Quality Standards to protect public health. The EPA is required to review these standards when considering new scientific evidence, and they have a team of staff and external scientists who periodically compile the scientific literature into an Integrated Science Assessment (previously called "Criteria Documents"). This literature is synthesized into a Staff Paper. The Integrated Science Assessment and Staff Paper are presented to and reviewed by the Clean Air Science Advisory Committee (CASAC), a panel of outside scientists and public health officials, in public hearings. CASAC then makes a recommendation to the EPA administrator, and the administrator publishes a decision in the *Federal Register*, along with a justification.

The EPA's review process often focused on the strengths and weaknesses of the studies under consideration. In early iterations of the review process, the relevant literature was winnowed to define the most informative and reliable studies. For example, in the 1987 promulgation of a new air quality standard for inhalable particles (PM_{10}),[11] only two epidemiologic studies of people exposed to measured air pollution were cited—the Steubenville Alert study[12] and a similar study of change in lung function of children following air pollution episodes in the Netherlands.[13]

As the number of air pollution health studies grew—particularly epidemiologic studies—it became clear that identifying key studies by elimination was not the best use of existing data or the reviewers' time. Based on recommendations from CASAC, the EPA turned to a weight-of-the-evidence approach. This approach was based on the landmark Bradford Hill criteria and was a broad, inclusive effort to examine the larger body of research. Reviewers carefully and comprehensively evaluated available evidence to see whether research results clustered around a consensus understanding.

The developing weight-of-the-evidence approach was challenged in the mid-1990s in the debate regarding new standards for $PM_{2.5}$. Although EPA Integrated Science Assessment documents[14] typically cite hundreds of studies, some opponents have attempted to frame the discussion around only a few studies. The expectation was that if they could undermine confidence in these studies, the basis for the $PM_{2.5}$ standard would fall.[15] Studies under heavy attack have included the children's respiratory hospital admissions study in Utah Valley,[16] the daily mortality time-series study from Philadelphia,[17] and the Harvard Six-Cities and ACS CPS-II cohort studies.[18]

SYSTEMATIC REVIEWS

The air pollution epidemiologic literature has grown dramatically over the years. The systematic review is growing with it, a complementary approach to evaluating the evidence. This approach emphasizes Hill's criteria and advances efforts to assess the consistency and coherency of evidence across multiple locations, populations, investigators, and health end points.

For example, in the early 1990s, Doug attended a meeting in Vancouver, where David Bates argued for air pollution researchers to focus on consistency and coherency (or overall cohesiveness) in the epidemiological literature. David argued that if air pollution was causally associated with increased cardiopulmonary mortality, it must also be associated with hospital admissions, emergency department visits, general practitioner visits, and other related health end points.[19] It stood to reason that other health effects might be caused by exposure to particle pollution, such as asthma or bronchitis attacks, elevated respiratory symptoms, more asthma medication, and reduced lung function.

Over the years, David's arguments influenced Doug and Arden, and they collaborated on various systematic reviews of the air pollution epidemiologic literature. These reviews observed reasonably consistent and coherent patterns related to respiratory and cardiovascular disease.[20] Further, a 2020 systematic review of air pollution and cancer provided consistent, cohesive evidence that exposure to $PM_{2.5}$ air pollution contributed to lung cancer.[21] Such reviews have been informative in identifying variations in observed associations and have led to deeper investigations.

SEARCH AND CONTROL FOR POTENTIAL CONFOUNDERS

Addressing the issue of potential confounding is critical in conducting studies of the health effects of air pollution and systematically reviewing these studies. A potential *confounder* would be a risk factor that causally affects health and is correlated with air pollution. So, for example, if air pollution does not causally affect health but is correlated with a risk factor that does, that other risk factor could be a confounder. A spurious (non-causal) association between air pollution and health effects would be observed if researchers did not control for or adequately statistically adjust for the confounder.

The potential for confounding is a major challenge in conducting and systematically reviewing studies of the health effects of air pollution. One obvious way to address this challenge is to search and control for potential confounders. Researchers ask the question: If it's not air pollution causing this association, what is it? What other variables are correlated with air pollution and could be causally affecting health? Can we use statistical models to estimate or isolate genuine pollution-health associations while adjusting for potential confounders?

A search for potential confounders could start with the early studies of severe air pollution episodes in Meuse Valley, Donora, or London, as discussed in chapter 2. Significant increases in respiratory and cardiovascular disease and death were associated with several days of extremely high air pollution concentrations. Was air pollution the causal factor, or was it something else? If air pollution was not the factor that caused the increase in respiratory and cardiovascular disease, what was?

Although all variables that potentially could impact health cannot be measured—that is, there were unmeasured covariates—there was no obvious or apparent confounder. There was, however, an essential concern regarding confounding in these studies. It was challenging to determine the causal effects of any specific air pollutant. Complex mixtures of contaminants from multiple sources were elevated in the unhealthy, polluted air—making it difficult to determine the impact of any specific pollutant or source.

As discussed in chapter 3, day-to-day variability in exposure to air pollution has been studied in hundreds of cities. The evidence is compelling. $PM_{2.5}$ air pollution is significantly associated with daily respiratory and cardiovascular mortality, hospitalizations, and adverse health outcomes. But is this association causal or is it due to confounding?

These studies are likely not confounded by cigarette smoking or socioeconomic or other factors without day-to-day changes linked with air pollution. The most likely potential confounders would be time-dependent variables such as weather, seasonality, long-term time trends, or day-of-week variables. However, pollution-health associations are observed in many cities with different weather conditions. The associations are observed even with various rigorous statistical approaches that adjust for these potentially confounding variables. And, as discussed in chapter 3, the innovation of the case-crossover methodology allowed for analyses that control for many potential confounders primarily by design rather than by statistical modeling. The adverse associations between air pollution and respiratory and cardiovascular health remained.

As discussed in chapter 4, many epidemiological studies use spatial or cross-sectional differences in long-term average exposure to air pollution. For example, early studies found evidence that mortality rates across U.S. metropolitan areas were associated with $PM_{2.5}$ air pollution. Even with statistical control for population-average differences in smoking rates and for socioeconomic, demographic, and other variables, there was no obvious confounder.

The prospective cohort studies discussed in chapters 4 and 5 also exploited spatial differences in air pollution. They further observed that long-term air pollution is associated with increased cardiopulmonary and lung cancer mortality risk. These pollution-mortality associations were observed after controlling for various potential confounders, including individual risk factors and community-level variables (contextual covariates). For example, the ACS CPS-II cohort analysis controlled for individual risk factors such as age, sex, race, smoking, alcohol use, marital status, education, body mass index, occupational exposures, and diet. Further analyses included smoothing out spatial differences and adjusting for contextual socioeconomic neighborhood variables. Again, the observed air pollution and mortality associations were not explained by any apparent confounders.[22]

An overall evaluation of episode, time-series, case-crossover, cross-sectional, and cohort studies does not provide evidence of a single or common confounder—not weather variables, not socioeconomic or demographic factors, not cigarette smoking. A compelling finding from a comprehensive evaluation of the air pollution and health literature is the lack of a single obvious confounder. Yet the lack of a confounder across epidemiological studies does not guarantee the elimination of all residual or unobserved confounding, nor does it ensure that observed associations are causal.

It's possible that air pollution and health studies have experienced "epidemiological bad luck."[23] It's *possible* that multiple confounders are synchronized across the different study designs, study areas, and study periods, resulting in pernicious but spurious associations. But this is unlikely. For this synchronization to occur, the confounders would need to be correlated with pollution across time and space, and they would have to be more potent risk factors for respiratory and cardiovascular than other diseases. As Doug, Arden, Rick Burnett, and others have surmised, the most likely explanation for the epidemiological evidence is that $PM_{2.5}$ and related air pollution actually do have observable and measurable effects on respiratory and cardiovascular health.[24]

RANDOMIZED CONTROLLED STUDIES

A conceptually and methodologically straightforward—and almost magical—solution can be used to meet research objectives when designing studies that rigorously try to identify causal relationships. This solution is *randomized controlled trials (RCT)*.

Randomized controlled trials are considered the best study design for determining the efficacy, or causal influence, of exposure to something because randomization typically balances potential confounders between the treatment and control groups. This design "simplifies the world" by randomly assigning research participants into well-defined experimental and control groups. RCT eliminates systematic differences across treatment groups and minimizes the potential for bias and confounding. RCT may be "blinded," using study protocols that prevent the researchers and participants from knowing which interventions are being received.

Randomized controlled trials are well-suited for evaluating the effectiveness of drugs, surgical procedures, and other medical treatments, and have been widely used for this purpose. Unfortunately, for ethical and pragmatic reasons, RCTs of air pollution are largely impractical, especially for long-term exposures. Therefore, epidemiological studies of air pollution rely primarily on efforts to glean information from exposure variability that is observable in real-world settings.

Randomized design does not control risk factors in the real world. Because some of these risk factors may be correlated with air pollution, there is the potential for confounding. Thoughtful researchers are cautious about making causal inferences while trying to explore and understand what knowledge can be gleaned from observational data. They also take advantage of opportunities to use natural experiments, quasi-experiments, and formal causal modeling approaches. These approaches are often designed to mimic the properties of randomized controlled trials.

NATURAL EXPERIMENTS, QUASI-EXPERIMENTAL STUDIES, AND CAUSAL MODELING

To understand how research can help identify causal relationships, it helps to understand research designs. There is no strict universal definition for study designs that enhance the ability to make causal inferences using epidemiological observational data. However, such designs often take advantage of opportunities to mimic RCT. They include natural experiments, quasi-experimental studies, and causal modeling approaches. Natural or quasi-experimental studies are studies where differences in exposure are not experimentally assigned by a researcher but are well-defined and determined by nature, public policy, the closing of a pollution source, or some other occurrence.

In studying real-world observational data, there is no single magic solution to establishing causality. However, the terminology, especially when referring to "causal models," can be misleading. Some of the most compelling evidence for causality can come from simple natural experiments or elegant, well-designed epidemiological studies using relatively simple or standard statistical analysis. Some excellent evidence of causality can come from skilled analysis of observational data using various "causal modeling" approaches. Yet using "causal modeling" approaches does not guarantee causality if not supported by adequate data or if the model's assumptions are not met.

There are a variety of strategies and approaches that enhance the ability to make causal inferences. The following are some examples.

Simple natural experiments. We have seen several examples of simple natural experiments in previous chapters. The studies in Meuse Valley, Donora, and London (chapter 2) were natural experiments where the elevated concentrations of air pollution during the episodes were not controlled experimentally but were well-defined in time and space and resulted from natural temperature inversions. Researchers did not randomly assign the study population exposures, but this wasn't necessary. The study populations were the same before, during, and after the episodes, so no randomization was required.

The Geneva Steel natural experiment was similar (chapter 3). The study did not evaluate the health effects in two populations. Instead, the study population was basically the same shortly before, during, and shortly after the mill's closure, requiring no randomization. The observed changes in air pollution were not researcher-controlled, but the intermittent operation of the steel mill provided significant differences in exposure.[25]

Interrupted time-series. In chapter 7, we discussed several complex natural or quasi-experimental studies. The analysis of these more complex quasi-experimental studies required data analysis that could be broadly defined as causal modeling. One such causal modeling approach is the *interrupted time-series (ITS)* design,[26] which analyzes data collected at equally spaced points in time before and after some "interruption"—perhaps a pandemic or a factory strike—to interpret the impact of the interruption. Examples of quasi-experimental ITS modeling approaches include the analyses of the sharp temporal changes that occurred due to the Dublin coal ban[27] and the copper smelter strike,[28] as discussed in chapter 7. This ITS approach strengthened the interpretation that sharp reductions in pollution causally reduced mortality.

Regression discontinuity. Regression discontinuity is another design used to evaluate the causal effects of specific interventions or exposures.[29] The study of spatial discontinuities created by China's Huai River policy, discussed in chapter 7, is an excellent example of the regression discontinuity design.[30] The findings of discontinuous elevations in particulate air pollution and discontinuous reductions in life expectancy just north of the Huai River were remarkably informative. The rigorous use of regression discontinuity analysis strengthened the causal interpretation that elevated air pollution from coal combustion contributed to respiratory and cardiovascular disease mortality.

Difference-in-differences. Difference-in-differences is another quasi-experimental design[31] used in air pollution and health research. The study of differences in life expectancies relative to differences in air pollution between 1980 and 2000,[32] as discussed in chapter 7, is a key example. The difference-in-differences design allowed stable cross-sectional differences to be controlled for as if by design. The use of formal difference-in-difference quasi-experimental analysis strengthens a causal interpretation that reductions in air pollution contributed to increases in life expectancy.

Another analysis using the difference-in-differences design evaluated differences in air pollution due to regulatory changes in the U.S. Clean Air Act Amendments compared with mortality in the Medicare population. This analysis argued that it provided evidence of a causal association between pollution and risk of mortality.[33]

Nonequivalent groups or quasi-control. This quasi-experimental design is similar to a natural experiment study. The difference is that in addition to the naturally exposed research group, there is also a comparable control group that did not experience the exposure. For example, this approach was used to study the effects of air pollution on school absences in three Utah school districts that experienced high levels of air pollution during winter temperature inversions. A neighboring and similar Utah school district was used as a quasi-control. This quasi-control school district was at a higher elevation and thus did not experience the same temperature-inversion–related elevated pollution levels. This analysis provided evidence that air pollution contributed to school absences. However, even with the control school district, it was difficult to disentangle the true effects of air pollution from other factors.[34]

Instrumental variables. Over the last few decades, the quasi-experimental instrumental variables design has been one of econometricians' most popular causal modeling approaches. More recently, it has been increasingly used in medicine and epidemiology.[35] Although formal mathematical demonstrations of the effectiveness of this approach can be complex, the basic design of the approach is reasonably straightforward and can be used effectively when adequate data are available.

This design requires identifying and using a special variable (the instrumental variable) that is related to or correlated with the treatment (or exposure of interest) but not to the outcome. So, for example, for a study of air pollution and mortality, researchers would use a variable associated with air pollution but not with mortality. Several air pollution and health studies have used the instrumental variable approach. For example, Schwartz, Bind, and Koutrakis conducted a daily time-series study of air pollution and mortality that used planetary boundary layer and wind speed as the instrumental variables (associated with air pollution but not with mortality). Using the appropriate methods to solve the model, they concluded that they provided evidence of a causal association between air pollution and daily deaths.[36]

Schwartz, Fong, and Zanobetti[37] later conducted an extensive national multi-city analysis using the instrumental variable design. Planetary boundary layer, wind speed, and air pressure served as instrumental variables. Again, the conclusion was that air pollution was causally associated with daily deaths. Interestingly, the pollution-mortality association was more prominent when using the quasi-experimental instrumental variables approach than a conventional time-series approach. Deryugina and colleagues conducted another interesting study. They used wind direction as an instrumental variable and estimated mortality and medical costs of air pollution among the U.S. elderly using Medicare.[38] In this causal model, mortality was linked with air pollution with significant life-years lost.

Propensity-score-based statistical methods. Various other statistical approaches have been developed to provide more robust causal inferences from observational data. Admittedly, these models require sophisticated mathematical and statistical modeling. However, these approaches can be simply understood as efforts to analyze nonrandomized observational real-world data in ways that attempt to mimic or approximate characteristics of a randomized controlled study.[39] Various related methods are sometimes used together to check the robustness of the results. These related approaches include propensity score matching, inverse probability weighting, covariate adjustment, and doubly robust regression.

Various air pollution studies have utilized these methods. For example, the Canadian cohort study of adults who moved to areas with different levels of air pollution (discussed in chapter 7) used propensity score matching to strengthen causal inference.[40] Another study of air pollution using data from Massachusetts employed a related causal modeling approach, generalized propensity score adjustment, and reported evidence that air pollution was "causally associated with mortality, even at levels below national standards."[41]

Several studies have used these causal modeling techniques with different cohorts of U.S. Medicare beneficiaries. One study used a cohort of approximately 13 million elderly adults who lived in the southeastern United States. They used doubly robust additive hazards models and observed additive causal associations between $PM_{2.5}$ air pollution and mortality.[42] Another analysis of approximately 17 million U.S. Medicare beneficiaries used inverse probability-weighted analysis to estimate the risk of death at each year of age. Again, $PM_{2.5}$ air pollution was associated with an increased risk of mortality.[43]

Perhaps the most remarkable use of propensity score-based causal modeling techniques was the huge nationwide U.S. Medicare cohort study discussed in chapter 5. Remember, it was partially motivated by the "secret science" debate. Harvard researchers constructed a massive Medicare cohort from data publicly available from the Centers for Medicare and Medical Services. They conducted standard or traditional statistical survival analysis.

As illustrated in figure 5.1, air pollution, especially $PM_{2.5}$, was strongly associated with elevated mortality risk.[44] To fully explore a causal interpretation of the $PM_{2.5}$-mortality association, the researchers reconducted the analysis using five different approaches. These approaches include the original Cox proportional hazard regression and Poisson regression approaches. They also used three causal inference approaches: a generalized propensity score matching approach, a generalized propensity score inverse weighting approach, and a generalized propensity score adjustment approach.[45] The estimated relationships between $PM_{2.5}$ and mortality risk were nearly the same across all five modeling approaches, including the causal modeling approaches.

The Harvard researchers made an extra effort to evaluate causal inference using the U.S. nationwide Medicare cohort.[46] They used propensity score inverse weighting with even more flexible modeling. The estimated response relationship between $PM_{2.5}$ and the relative mortality risk from this analysis is similar to the relationship represented in figure 5.1. A troubling finding was the easy-to-observe result that pollution-mortality effects occurred even at levels well below the National Ambient Air Quality Standards. The findings of nearly identical $PM_{2.5}$-mortality associations with multiple standard regression and causal modeling approaches in the nationwide Medicare cohort supported a causal inference regarding the air pollution-mortality relationship.

Final illustrative example. As a last example of causal inference modeling, consider another cohort study of air pollution and mortality motivated by the "secret science" debate. This study used twenty-nine years of U.S. National Health Interview Survey (NHIS) data. As reported in chapter 5 and as illustrated in figure 5.2, $PM_{2.5}$ air pollution was associated with an increased mortality risk, especially cardiopulmonary and lung cancer mortality.[47]

The results of the NHIS cohort studies seemed compelling, partly because the NHIS survey data were collected to be a large representative sample of the U.S. population. So, what concern could there be? Regarding exposure to air pollution, there was no experimental randomization. Long-term ambient air pollution exposures were determined by where people lived. What if there was systematic, nonrandom sorting? What if persons with characteristics that increased their risk of death (especially for cardiopulmonary and lung cancer) tended to live in more polluted areas? For example, what if more polluted areas consisted of relatively poorer people, older people, or people who were more likely to smoke? If such nonrandom sorting existed, observed associations between air pollution and mortality risk could be due to confounding of these other correlated risk factors.

The good news is that the NHIS survey also collected data on individual risk factors, including age, sex, race and ethnicity, smoking status, income levels, education levels, marital status, and body mass index. In the team's initially reported results, they controlled for these risk factors using statistical models (multivariate Cox proportional hazard regression models). Pollution associations with increased mortality risk, especially cardiopulmonary and lung cancer mortality, persisted even when controlling these other risk factors.[48] However, the team also wanted to analyze the data using formal causal modeling approaches. Their first preference was to use an instrumental variable approach, but they were unable to find a viable instrumental variable.

Ultimately, to further explore a causal interpretation of the $PM_{2.5}$-mortality association, the team analyzed the data using inverse probability weighting.[49] This approach weighted the data to mimic or approximate the characteristics of a randomized controlled study. In effect, this approach weights or adjusts the data such that the results that would occur were as if the NHIS cohort had been randomized across pollution levels relative to the observed risk factors. The team used this approach without controlling the other measured risk factors and also with control of these factors. Inverse weighting with control of the measured risk factors resulted in a dual or double control for confounding, becoming a "doubly robust" analysis. The team also evaluated the robustness of the study using various distributional assumptions.

What were the results, and was there evidence of confounding? Long-term $PM_{2.5}$ exposures were associated with increased all-cause and cardiopulmonary mortality. The associations for all-cause mortality were nearly the same as was estimated with standard Cox proportional hazards models. Like other studies, the $PM_{2.5}$-mortality associations were strongest for cardiopulmonary mortality. However, the estimated associations were nearly identical across standard regression or causal modeling approaches. There is always some risk of residual confounding by some unknown, unmeasured confounder. Yet these results suggested minimal potential confounding due to measured variables and provided further evidence supporting a causal inference.

Making causal inferences using observational data is challenging. Some quasi-experimental designs and analytic methods are complex and difficult to explain in a short chapter. However, a substantial and impressive body of epidemiological evidence suggests that air pollution contributes to respiratory and cardiovascular disease.

CONCLUSION

"Does air pollution research provide evidence of a causal relationship?"

There is no single magic solution to proving causality. Still, the body of epidemiological evidence is compelling. Air pollution contributes to respiratory and cardiovascular disease and death—causally.

Some would dismiss this body of evidence, arguing that "correlation isn't causation" and suggesting that air pollution and health researchers are naive and unaware of the challenges of making causal inferences from observational data. But this is generally not true. There is a massive amount of epidemiological research, and it has been evaluated using weight-of-the-evidence and systematic-review approaches. Researchers made concerted efforts to seek and control or statistically adjust for potential confounding. Furthermore, much of this research took advantage of remarkable natural experiments, quasi-experimental studies, and causal modeling approaches to evaluate causal inference rigorously.

So, what would provide further proof? More and more high-quality quasi-experimental studies may help. It would also be helpful to know if the epidemiological observed health effects of air pollution are biologically plausible.

10

HOW DOES AIR POLLUTION CAUSE HEALTH EFFECTS?

The evidence is compelling. Air pollution contributes to respiratory, cardiovascular, and lung cancer disease and death. But just how does it cause these health effects?

It's crucial to study the biological pathways, or "mechanisms," that link air pollution with these adverse health effects. Understanding these mechanisms confirms observational studies, provides insights into biological science, and identifies potential medical interventions and treatments. Learning *how* pollution adversely affects health supports the conclusion *that* pollution affects health.

It is not surprising, then, that there has been an ongoing interest in an improved understanding of the mechanistic pathways of the effects of $PM_{2.5}$. In an editorial in *Science* titled "A Call for More Science in EPA Regulations," William. J. Madia, director of the U.S. Department of Energy Pacific Northwest National Laboratory, called for "a more credible, scientific basis for environmental regulation." He suggested that air pollution policy should rely on research "directed at understanding the underlying biological mechanisms by which the population experiences health effects."[1]

Arden and Doug collaborated on studies that explore biological mechanisms of pollution-health effects. As with epidemiological evidence, the overall evidence of biological mechanisms comes from many collaborating research teams.

MECHANISTIC PATHWAYS

What biological mechanisms explain why breathing polluted air adversely affects respiratory and cardiovascular health?

Biological mechanisms include systems of causally interacting processes and pathways that result in the observed health effects. An example would be the connections between the lungs, heart, and circulatory system we've discussed in previous chapters. Multiple mechanisms interact with each other in complex biological systems. Even a partial or general understanding of the biological mechanisms or physiological pathways that link $PM_{2.5}$ air pollution with respiratory and cardiovascular disease and death would help. Knowledge of these mechanistic pathways could strengthen fundamental inferences regarding causality and help researchers develop medical or other strategies to mitigate the effects of air pollution.

Detractors of the evidence that air pollution adversely affects cardiopulmonary health have argued that these effects are not biologically plausible. In his 1997 critical review, "Ambient Particles and Health: Lines That Divide," pulmonologist Sverre Vedal perceptively noted that there was some truth to this concern. He indicated that a key "stumbling block" was "weak biological plausibility" or "ignorance or inadequate understanding of biological mechanisms."[2]

Much changed, however, during the ten years following that article's publication. Arden and Doug's 2006 critical review, "Health Effects of Fine Particulate Air Pollution: Lines That Connect,"[3] highlighted research progress on the mechanistic pathways that connected $PM_{2.5}$ exposure with cardiopulmonary disease and mortality. It also presented a stylized illustration of general physiological pathways.

Over the subsequent years, there has been much additional research regarding biological mechanisms. Figure 10.1 provides a revised and updated illustration of general mechanistic pathways regarding air pollution and cardiopulmonary health. Figure 10.1 is not exhaustive, nor does it present the enormously complex biological, mechanistic details

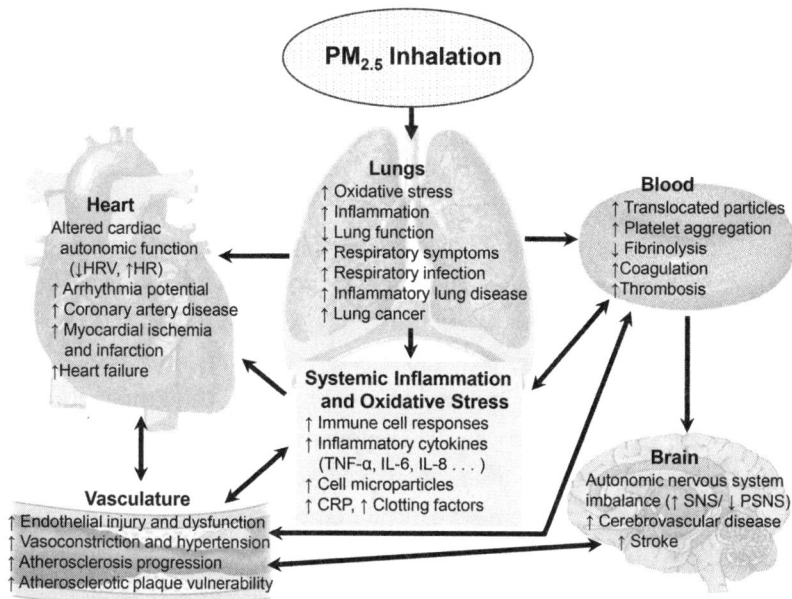

Figure 10.1
Stylized illustration of general pathophysiological mechanistic pathways linking PM$_{2.5}$ air pollution exposure with cardiopulmonary disease. *Source:* Modified and updated from C. Arden Pope and Douglas W. Dockery, "Health Effects of Fine Particulate Air Pollution: Lines That Connect," *Journal of the Air & Waste Management Association* 56, no. 6 (2006): 709–742.

involved. It illustrates that, based on the evidence, air pollution affects more than just the lungs. Mechanistic pathways involve the body as a whole, including systemic inflammation and oxidative stress, which impact the lungs, blood, vasculature, heart, brain, and autonomic nervous system.

IT STARTS IN THE LUNGS

The primary function of the lungs is to send oxygen from inhaled air into the red blood cells in our systemic circulation. Simultaneously, they return carbon dioxide from these same cells into the exhaled air. In both directions, this happens through a process called *passive diffusion*. Healthy lungs easily transport oxygen into the bloodstream and remove carbon dioxide. This amazing feat is accomplished by presenting a large surface area comprised of extraordinarily fragile and thin tissue in the alveolar gas exchange regions of the lung. Although it is difficult to measure precisely, the gas exchange surface area in healthy adults is estimated to be between 70 and 145 m^2, almost half the size of a singles tennis court.[4] Fine particles or other contaminants in the ambient air that get past the respiratory defense mechanisms to this gas exchange region can damage the lungs or gain access to the systemic circulation and other organ systems.

It has long been known that breathing air polluted with $PM_{2.5}$ can adversely affect the lungs. In 2000, Peter Gehr and Joachim Heyder edited a 700-plus-page book that addressed "particle-lung interactions."[5] Over twenty years later, any book comprehensively covering the body of research on particle-lung interactions and related health effects would be much longer. Since then, several pivotal mechanisms for the adverse effects of $PM_{2.5}$ air pollution have emerged.

Let's take it step by step.

LUNG FUNCTION

Inhaled $PM_{2.5}$ air pollution, as foreign matter in the lungs, is implicated in eliciting pulmonary inflammation, respiratory symptoms, and decrements in lung function. For example, the early episodic studies discussed in chapter 2 provided evidence that breathing polluted air resulted in significant inflammation in the airways of the lungs of animals and humans. This pollution led to an increase in coughing, breathing difficulties, and symptoms of cardiovascular disease. As noted in chapter 3, in controlled toxicological experiments in Utah Valley, particulate matter air pollution extracted from the archived air monitoring filters elicited acute airway injury and inflammation in rats and humans.[6]

Doug, with collaborators Ben Ferris, Frank Speizer, and others, conducted early studies (1980s and 1990s) on particulate air pollution and its impact on lung function and respiratory symptoms. These studies were part of the Harvard Six-Cities research project.[7] Although the initial focus of the Harvard Six-Cities study was the chronic respiratory effects of long-term air pollution exposure, it became apparent that there was an opportunity, indeed a need, to examine *acute* effects.

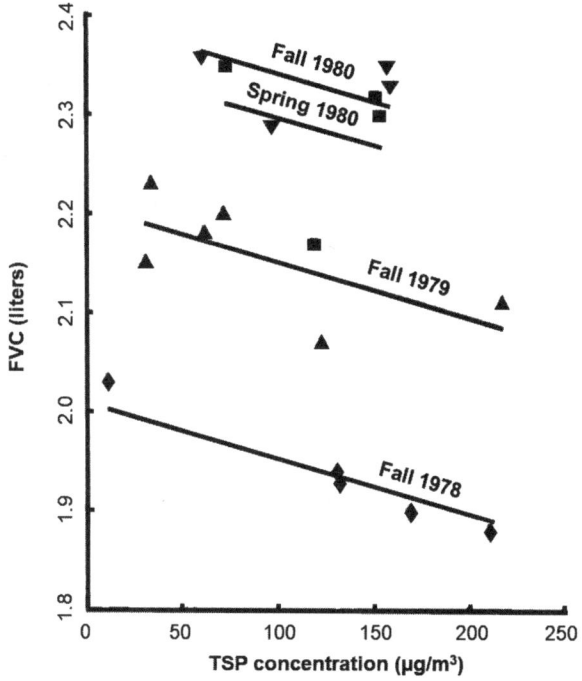

Figure 10.2

Lung function (forced vital capacity, FVC) versus total suspended particulate (TSP) concentration for a sample child who participated in all four Steubenville alert studies. *Source:* Douglas W. Dockery, James H. Ware, Benjamin G. Ferris, Frank E. Speizer, Nancy R. Cook, and Stanislaw M. Herman, "Change in Pulmonary Function in Children Associated with Air Pollution Episodes," *Journal of the Air Pollution Control Association* 32, no. 9 (1982): 937–942. Enhanced imaged reprinted by permission of the publisher (Taylor & Francis Ltd.).

High air pollution events (alerts) were regularly observed each fall in Steubenville, the most polluted of the six cities. In a four-year study, a subsample of third- and fourth-grade schoolchildren received baseline lung function tests early in the fall. During each air pollution alert, these children were tested. After each alert, they were retested one, two, and three weeks later. The study design assumed a single high exposure during the air pollution alert with relatively clean air at the other observation points. But the reality was that air pollution followed a continuum throughout the study periods, with air pollution even on some of the "clean" days approaching the alert level.

The team analyzed the combined data across all study periods using repeated measures of lung function and daily air pollution. A total of 330 children had three or more measurements, and 194 participated in more than one of the panel studies. Lung function declined on days with higher particulate air pollution.[8] Figure 10.2 illustrates lung function (forced vital capacity, FVC) versus total suspended particulate (TSP) concentration for a child who participated in all four Steubenville alert studies. The team had assumed that lung function would change over weeks. However, the analyses of the multiple episodes showed very quick change—on the same or succeeding day of exposure.[9]

Doug continued his collaborative efforts to explore the effects of air pollution on lung function and respiratory diseases. These efforts included the panel studies discussed in chapter 3, conducted with Arden in Utah Valley;[10] another study in Uniontown, Pennsylvania;[11] and multi-city studies of long-term exposures.[12] Other studies by various research teams, by Joel Schwartz,[13] and a series of large multicenter studies in Europe have further confirmed that air pollution contributes to deficits in lung function and respiratory illness.[14]

The Southern California Children's Health Study, discussed in chapter 7, provided compelling evidence that long-term exposure to air pollution contributed to deficits in lung function and increased respiratory symptoms and disease.[15] There is related evidence that air pollution increases the risk or the need for healthcare for acute lower-respiratory infections, including influenza, respiratory syncytial virus (RSV), and serious colds and pneumonia.[16]

OXIDATIVE STRESS IN LUNGS

In 1991, researchers introduced a relatively new biology term—oxidative stress—during early research on air pollution, lung function, and inflammatory lung disease.[17] Much subsequent research has suggested that mechanisms related to oxidative stress and inflammation may play a central role in many of air pollution's health effects. Various air pollutants, including $PM_{2.5}$ and ozone, are potent oxidants or can generate *reactive oxygen species* (ROS) that contribute to oxidative stress in the lungs. Fine particulate matter air pollution ($PM_{2.5}$) contains metals and organic chemicals that contribute to the generation of ROS. Tiny airborne particles have large surface areas that can transport free radicals (highly unstable and reactive atoms or molecules that can damage cells) and support reactions in the lungs, thus contributing to oxidative stress.

Oxidative stress occurs with an imbalance of ROS relative to antioxidant defenses and can lead to cell and lung tissue damage. Given the large surface area of the lungs and their direct exposure to air pollution, the lungs are a primary target organ for oxidative injury.[18]

Within the lungs, oxidative stress is linked to pulmonary inflammation and diseases such as asthma, chronic obstructive pulmonary disease (COPD), and pneumonia. A critical response to oxidative stress–related lung damage is increased inflammatory cells. Given that activated inflammatory cells also generate and release free radicals, which can further damage cells, this response leads to a second wave or cascade of oxidative stress and inflammation in the lung.[19] Oxidative stress and related pulmonary inflammation, therefore, are part of the general mechanistic pathway through which air pollution contributes to inflammatory lung disease, including respiratory symptoms and adverse effects on lung function.[20]

LUNG CANCER

As discussed in previous chapters, air pollution is strongly linked to another critical disease: lung cancer. The International Agency for Research on Cancer (IARC) classified particulate matter air pollution as a human carcinogen (Group I) for lung cancer.[21] Evidence that $PM_{2.5}$ air pollution contributes to lung cancer continues to grow.[22] The mechanistic pathways that explain the relationship between air pollution and lung cancer are not fully understood. Even so, particulate air pollution contains many mutagens and carcinogens.

Further, as discussed before, air pollution contributes to oxidative stress. ROS, oxidative stress, and inflammation likely play a role in the mechanisms related to the development of lung cancer.[23] To this point, a remarkable study of patients who had heart disease and had experienced a heart attack showed that anti-inflammatory therapy "could significantly reduce incident lung cancer and lung cancer mortality."[24]

In September 2022, scientists from the Francis Crick Institute and University College London presented evidence at the European Society for Medical Oncology (ESMO) Congress in Paris.[25] This team had researched mutations seen in a specific gene (EGFR) in about half of nonsmoking people. The researchers found that increased exposure to $PM_{2.5}$ air pollution was linked with an increased risk of lung cancer among these nonsmokers. Laboratory studies demonstrated the potential mechanistic roles of air pollution-related inflammation in lung cancer. William Hill, a member of the team, said, "Finding ways to block or reduce inflammation caused by air pollution would go a long way to reducing the risk of lung cancer."[26]

THE AUTONOMIC NERVOUS SYSTEM AND CARDIAC FUNCTION

The late 1990s were enlightening years for Arden, Doug, and other air pollution researchers. Earlier in the decade, they and others conducted research exploring the effects of air pollution on lung function and respiratory illness. It was presumed that the health effects of air pollution were confined to impacts on the lungs and respiratory system.

However, one of the big surprises from their research was the finding from the time-series studies (discussed in chapter 3) that elevated air pollution increased not only respiratory hospitalizations and deaths but also hospitalizations and deaths due to *cardiovascular* disease. In the mid-1990s, they published the initial findings of the Harvard Six-Cities and the ACS CPS-II cohort studies (see chapter 4), which indicated that air pollution had significant effects on cardiopulmonary disease that included both respiratory and cardiovascular disease mortality. In the late 1990s (see chapter 4), these two cohort studies were involved in intense debate and the HEI independent reanalysis. At this time, Doug and Arden were busy trying to understand the mechanistic pathways that could explain observed air pollution-related increases in cardiovascular disease.

Doug and Arden initially thought a physiological pathway might include declines in blood oxygenation with elevated exposure to air pollution, which would increase stress on the heart. During the winter of

1995–1996, they studied a panel of ninety retired adult subjects in Utah Valley. They hypothesized that the effect of air pollution on oxygen saturation could be more easily observed in a relatively high-elevation mountain valley. They measured subjects' blood oxygen saturation daily using pulse oximetry (a simple, painless test that measures the oxygen levels of the blood and the pulse rate).

As hoped, this panel study design had remarkable statistical power. It observed the expected but minimal positive associations between blood oxygen saturation and atmospheric barometric pressure. But contrary to their prior hypothesis, elevated levels of particulate air pollution were *not* significantly associated with decreased blood oxygen saturation.[27]

This result was disappointing and puzzling. However, Doug and Arden received help with their dilemma from a colleague, John Godleski, a professor of pathology at Harvard Medical School. John was conducting innovative controlled exposure crossover experimental studies on the effects of air pollution using dogs. He found that exposure to particulate air pollution induced changes in cardiac rhythm, specifically, mean heart rate and pulse variability.[28] John suggested they could go back and evaluate air pollution and pulse rate because the pulse oximeters also measured heart rate.

Doug and Arden did as John suggested. They found that pulse rate and its odds of substantial elevation (by 5-plus or 10-plus beats per minute) were associated with high levels of particulate air pollution.[29] These results suggested a pathway related to cardiac rhythm and cardiac autonomic function rather than oxygen saturation.

With these results in mind, Arden and Doug conducted additional research on air pollution and cardiac rhythm. Specifically, they evaluated air pollution exposure associated with *heart rate variability (HRV)* measures. HRV reflects the heart's ability to adapt automatically to internal or external stimuli. HRV involves complex *autonomic nervous system (ANS)* interactions between the brain and cardiovascular system and provides specific measures of cardiac autonomic function. Reduced HRV is a marker of compromised cardiac autonomic function and greater cardiac vulnerability.

To further explore this alternative pathway, Arden and Doug designed a pilot study to assess the impacts of air pollution exposure on people in their daily lives, using a rigorous and direct measure of *electrocardiograph (ECG)* changes. They borrowed several portable ambulatory ECG Holter monitors from a collaborator, Mark Raizenne. They recruited six participants from the oximetry study and one of the field technicians to monitor their heart continuously for twenty-four to forty-eight hours on up to three occasions (see figure 10.3).[30] They measured ECGs on each subject before, during, and after air pollution episodes in Utah Valley.

After the team began the ECG monitoring, PM_{10} concentrations rose and fell during the winter. Thus, electrocardiographic data were collected over a wide range of PM_{10} exposures for each subject. This small pilot study showed that particulate air pollution was associated with reduced heart rate variability, suggesting that air pollution had an unhealthy effect on cardiac autonomic function in otherwise healthy, older subjects.[31]

Could these results be accurate? Arden and Doug engaged in several separate but similar studies to answer this question. Arden and colleagues followed up the pilot study by conducting a panel study of eighty-eight subjects, using twenty-four-hour ambulatory ECG monitoring. The new

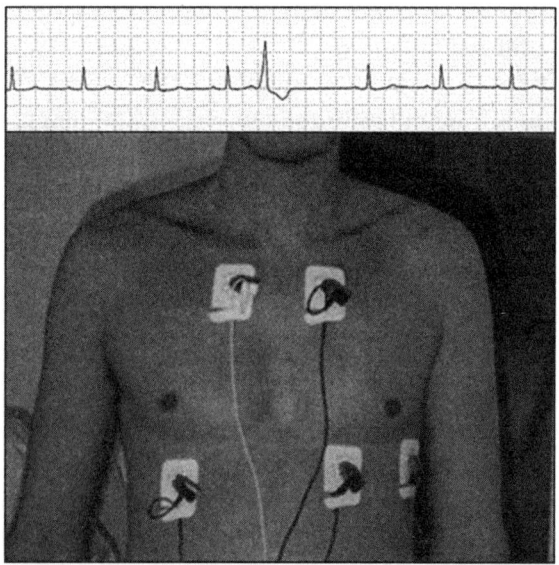

Figure 10.3
Research subject being prepared for ambulatory ECG monitoring with a small sample of heartbeat recording provided above.

study confirmed that elevated air pollution exposure was associated with increased heart rate, decreased heart rate variability, and compromised cardiac autonomic function.[32] However, the researchers observed no associations between air pollution and other measures related to cardiac function, including ventricular repolarization[33] and atrial fibrillation hospitalization.[34]

Doug, meanwhile, led a related but novel study of air pollution and cardiac arrhythmias. Patients with arrhythmias often have an *implanted cardioverter defibrillator (ICD)* that continuously monitors irregular heartbeats and measures heart function. On detecting an abnormality, the ICD can intervene by delivering electric shocks to restore proper electrical function. These devices document the type and time of these potentially serious irregular heartbeats.

Doug reached out to Mark Link, a young cardiologist at the Cardiac Arrhythmia Center of Tufts Medical Center in Boston. They recruited a panel of 200 cardiac patients with ICDs and followed them from 1995 through 2002. The team found weak evidence that particulate air pollution was associated with triggering nonfatal ventricular arrhythmias.[35] Furthermore, unlike the analysis of the panel of older subjects in Utah noted above, this study found that elevated exposures to air pollution were significantly associated with an increased risk of atrial fibrillation.[36]

In June 1999, Arden gave a presentation about the health effects of air pollution at the International Society of Aerosols in Medicine conference in Vienna, Austria. This presentation discussed particulate air pollution, the autonomic nervous system, and heart rate variability.

Two interesting things happened while traveling home after the meeting. First, Arden had some extra time before the return flight. As someone who grew up working with and loving horses, Arden visited the Vienna Spanish Riding School to watch the training and performance of the famous Lipizzaner horses. The horses and trainers were amazing. But he stayed too long and missed his flight. Second, while in the airport waiting for the next available flight, it became apparent that the in-airport smoking restrictions were not well enforced. Arden was exposed to a noticeable amount of secondhand cigarette smoke. Thinking of his recent research and the presentation at the conference, he wondered if this cigarette smoke could affect his ANS and HRV.

Arden spent much of his long journey home from Vienna thinking about and outlining protocols for a study of HRV and secondhand smoke. Upon returning home, he contacted Delbert Eatough, a professor of chemistry and biochemistry at Brigham Young University. Delbert had substantial experience monitoring air pollution, including secondhand cigarette smoke on airplanes and elsewhere.

The team undertook a study in the Salt Lake City International Airport, which at the time had well-defined smoking areas within the otherwise smoke-free terminals. They evaluated changes in secondhand smoke exposures and HRV in nonsmoking adults using ambulatory ECG and portable air pollution monitoring equipment. The research participants, over eight-hour periods, alternated two hours in nonsmoking and smoking areas. Cardiac autonomic function, as measured by HRV, declined when participants were exposed to high concentrations of cigarette smoke. HRV then partially recovered when the subjects moved back to the nonsmoking areas.[37]

Many other studies since then have evaluated air pollution and HRV. One meta-analytic review of twenty-nine such studies concluded that elevated particulate air pollution was associated with declines in HRV.[38] In other words, the heart was less likely to be able to make normal, healthy adjustments to its rate.

Another interesting quasi-experimental study of workplace smoking bans observed that HRV improved for nonsmokers with the introduction of smoke-free workplaces.[39] These results suggested that exposure to particulate air pollution, including secondhand cigarette smoke, influenced the ANS and adversely affected cardiac autonomic function.

SYSTEMIC INFLAMMATION, OXIDATIVE STRESS, BLOOD

By the early 2000s, cardiovascular research had established that *systemic inflammation* was a significant contributor to *atherosclerosis*, a disease of the blood vessels. Arteries, the blood vessels that carry blood throughout the body, are lined with an inner layer of cells called the *endothelium*. Atherosclerosis involves the endothelium's buildup of fats, cholesterol, and other substances. This buildup results in plaques that narrow, constrict, and harden the arteries.

Atherosclerotic plaques can obstruct blood flow, rupture, and create a potentially deadly clot. Atherosclerosis is the underlying disease responsible for most adverse cardiovascular events, including heart attacks and strokes. It has become apparent that systemic inflammation contributes to endothelial damage, the development of atherosclerosis, and the acute complications of atherosclerosis.[40]

So, what does research on systemic inflammation have to do with the effects of breathing polluted air into the lungs? As researchers continued exploring the potential biological pathways that explained the observed health effects of air pollution, they learned that what starts in the lungs does not stay in the lungs. For example, particulate air pollution, especially ultrafine particles, can move from the lungs into the blood and cause inflammation in other body parts,[41] including the brain.[42]

Even more crucially, systemic, or widespread, oxidative stress and inflammation constitute mechanistic pathways resulting in pollution health effects beyond the lungs. Air pollution results not only in localized pulmonary inflammation and oxidative stress. It also causes a "systemic spillover," resulting in broad oxidative stress and inflammation.[43] Therefore, exposure to $PM_{2.5}$ air pollution results in an elevated marker of inflammation, *C-reactive protein (CRP)*. Air pollution exposure can trigger immune cell responses and the release of inflammatory cytokines (such as TNF-α, IL-6, and IL-8). Responses may also include elevated blood clotting (with an increase in clotting factors, coagulation, thrombosis, and platelet aggregation) and reduced breakdown of blood clots (fibrinolysis). Inflammatory endothelial damage could also cause elevated levels of endothelial microparticles into the blood.[44]

In 2003, previous collaborators on the ACS CPS-II cohort study teamed up with Harvard pathologist John Godleski to explore epidemiological evidence of disease pathways. The team focused on three hypothesized general physiological pathways, including (1) accelerated progression of respiratory disease, (2) systemic inflammation and accelerated atherosclerosis, and (3) altered cardiac autonomic function. The results were intriguing. It was expected that air pollution would be most adversely associated with chronic *respiratory* disease. But no. It was strongly associated with *cardiovascular mortality*. The results were limited by reliance on cause-of-death coding from death certificates and the existence of substantial comorbidity between respiratory and cardiovascular disease and death. Even so, the results were consistent with the hypothesis that the primary general physiological pathways linking exposure to $PM_{2.5}$ air pollution with cardiopulmonary mortality include pulmonary and systemic inflammation, endothelial or vascular injury, and accelerated atherosclerosis.[45]

Systemic oxidative stress and inflammation may be physiological pathways responsible for much respiratory and cardiovascular disease.[46] Chronic respiratory disease, demonstrated by respiratory symptoms or deficits in lung function, is also a substantial risk factor for cardiovascular disease and death.[47] But is there even more direct evidence?

BLOOD VESSELS, HEART, BRAIN

In the late 1990s and early 2000s, Stephan van Eeden, James Hogg, and colleagues at the iCAPTURE Centre for Cardiovascular and Pulmonary Research at the University of British Columbia conducted a series of pioneering studies. They found that exposure to particulate matter (ambient air pollution or cigarette smoke) contributed to pulmonary inflammation. Inflammation in the lungs spilled over to systemic inflammatory responses, promoting the development of atherosclerotic plaques. Studying rabbits prone to atherosclerosis, the researchers found that exposure to particulate pollution accelerated the progression of atherosclerosis and resulted in greater vulnerability to plaque rupture, which can lead to heart attacks or stroke.[48]

Another fascinating animal study (using mice) was conducted by a research team led by Qinghua Sun, with collaboration by prominent cardiologists and pioneers in environmental medicine, including Sanjay Rajagopalan, Robert Brook, Morton Lippmann, Lung-Chi Chen, and others.[49] Mice genetically predisposed to developing atherosclerosis were experimentally exposed to combinations of high-fat versus normal diets and clean, filtered air versus air moderately polluted with $PM_{2.5}$. Experimental pollution exposures, normalized over the six months of exposure, were comparable to common exposures observed in moderately polluted U.S. cities and much lower than exposure common in highly polluted cities such as in China or India. Over six months of exposure, polluted air induced vascular inflammation and exacerbated atherosclerosis. The effects of exposure to air pollution were statistically significant and could be observed in images (photomicrographs) of blood vessels, providing tantalizing evidence that exposure to $PM_{2.5}$ air pollution contributes to atherosclerosis. Combining $PM_{2.5}$ exposure with a high-fat diet produces even more diseased blood vessels.

At about the time Sun's mouse study was reported, Nino Künzli and colleagues at the University of Southern California published evidence that $PM_{2.5}$ air pollution was associated with atherosclerosis in humans. Using data of patients enrolled in clinical trials, they found that exposure to air pollution was associated with the thickness of the carotid artery's inner layers (carotid intima-media thickness, CIMT), a primary measure of subclinical atherosclerosis. (A "subclinical" condition does not cause a health problem big enough to lead to a disease diagnosis or death, but harms the body anyway.) The Künzli study provided the first direct epidemiological evidence of an association between air pollution and a subclinical measure of human atherosclerosis.[50]

In 2002, the EPA requested proposals to study the long-term cardiovascular consequences of chronic exposure to $PM_{2.5}$. The goal was to conduct a highly informative study of chronic $PM_{2.5}$ exposure effects on clinical measures of atherosclerotic disease. Many research teams applied, including teams that included Arden and Doug. Doug teamed up with Nino Künzli, and Arden teamed up with Robert Brook. Neither of their teams was selected.

The selected proposal came from a team led by Joel Kaufman and colleagues at the University of Washington. They proposed to study participants in the ongoing Multi-Ethnic Study of Atherosclerosis (MESA) clinical study. Research subjects from six states participated in a detailed clinical follow-up of atherosclerosis progression using coronary artery calcification and other advanced clinical measures. The new study was dubbed MESA-AIR.

Although Doug's and Arden's proposals were not selected for funding, they supported Joel Kaufman and colleagues' winning proposal. They thought it was excellent, with a strong study design and a high-quality research team. Doug served on the External Science Advisory Board of MESA-AIR from 2004 to 2014 (chair, 2004–2008), and Arden served on a project review committee. What did the study find?

Joel Kaufman and collaborators on the MESA-AIR study conducted multiple analyses showing that air pollution exposure contributed to cardiovascular disease. For example, $PM_{2.5}$ air pollution was associated with elevated blood pressure, especially near roadway traffic.[51] The researchers also reported that elevated exposures to $PM_{2.5}$ pollution were associated with increased progression of atherosclerosis (again based on CIMT).[52] Notably, they found that "increased concentrations of $PM_{2.5}$ and traffic-related air pollution within metropolitan areas, in ranges commonly encountered worldwide, are associated with progression in coronary calcification, consistent with acceleration of atherosclerosis."[53] They further

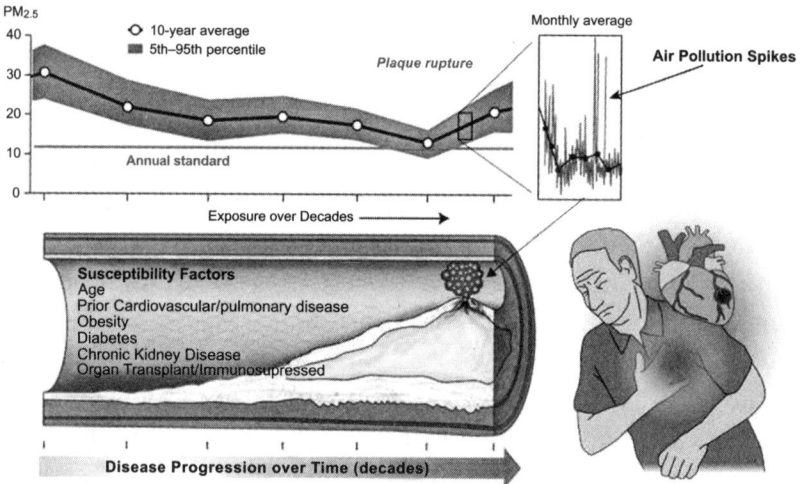

Figure 10.4

Illustration of $PM_{2.5}$-induced atherosclerosis. From left to right, long-term, chronic exposures contribute to the initiation and progression of vascular disease. A short-term period of elevated $PM_{2.5}$ exposure increases risk of plaque rupture and acute cardiovascular events such as heart attacks and strokes. *Source:* Graham H. Bevan, Sadeer G. Al-Kindi, Robert D. Brook, and Sanjay Rajagopalan, "Ambient Air Pollution and Atherosclerosis: Recent Updates," *Current Atherosclerosis Reports* 23, no. 10 (2021): 63. Reproduced with permission from Springer Nature.

concluded that their research "supports the case for global efforts of pollution reduction in prevention of cardiovascular diseases."[54]

Evidence continues to mount that long-term exposure to $PM_{2.5}$ air pollution contributes to the presence and progression of atherosclerosis in humans (see figure 10.4).[55] There is also growing evidence that air pollution can trigger acute cardiovascular events by adversely affecting atherosclerotic plaque vulnerability, likely through systemic inflammatory pathways.[56] Other studies have provided and evaluated evidence that air pollution contributes to vasoconstriction, hypertension,[57] and vascular and endothelial injury.[58]

THE INFLAMMATORY PATHWAY, SUMMARIZED

To recap, the following describes one general physiological pathway that links $PM_{2.5}$ air pollution with cardiopulmonary disease (see figure 10.1).

Fine particulate matter ($PM_{2.5}$) and related pollutants are breathed into the lungs. These pollutants initiate oxidative stress and inflammatory responses and contribute to increased respiratory symptoms, lung function deficits, inflammatory lung disease, and increased risk of lung cancer.

Oxidative stress and inflammation do not remain localized to the lungs. Rather, they become systemic, leading to immune cell responses, the release of inflammatory cytokines (such as TNF-α, IL-6, and IL-8), and other markers of inflammation (CRP). These responses also include elevated blood clotting (increase in clotting factors, coagulation, thrombosis, platelet aggregation) and poorer breakdown of blood clots (fibrinolysis). This inflammatory endothelial damage releases elevated endothelial microparticles into the blood, contributing to vascular disease, including atherosclerosis.

The vascular disease outlined above underlies much of the common coronary artery disease and cerebrovascular disease associated with air pollution. Chronic, long-term exposure to air pollution contributes to the initiation and progression of atherosclerosis and eventually leads to the development of vulnerable atherosclerotic plaques. Short-term acute elevated exposures increase the risk of triggering acute coronary or cerebrovascular events such as heart attacks or strokes.[59]

EPISODIC EXPOSURE AND SUBCLINICAL SYSTEMIC INFLAMMATION

Aruni Bhatnagar is a professor of medicine and distinguished university scholar at the University of Louisville. He specializes in environmental medicine, focusing on mechanistic pathways linking environmental exposures to cardiovascular disease. Around 2008, Aruni and Arden were in Seattle, Washington, serving on a project research review committee for the MESA-AIR study. One evening, Aruni and Arden went to dinner together and had a discussion that resulted in a fruitful research collaboration. Specifically, they would work together to explore whether episodic exposures to $PM_{2.5}$ air pollution were associated with systemic inflammation and injury to the endothelium, the cells lining the blood vessels.

In the resulting study, they collected samples from healthy, nonsmoking, young adults. Because they could not directly control the pollution levels in the air the subjects were breathing, this was not a controlled experimental design. But they could predict periods of relatively clean and polluted air by watching for local weather conditions that would or would not result in stagnant air conditions. Then they could control the timing of the blood draws such that they drew blood across periods of different levels of air pollution. Since all the research subjects were young and healthy, air pollution episodes were not expected to have clinically significant impacts on their health. However, the researchers evaluated the *subclinical* effects of air pollution.

Aruni and Arden hypothesized that episodic exposure to $PM_{2.5}$ air pollution would result in immune cell responses and increased endothelium-derived microparticles in the blood, markers of subclinical endothelial or vascular injury (damage to blood vessels). Teaming with Timothy O'Toole, professor of medicine at the University of Louisville and director of a flow cytometry lab, they measured microparticles and immune cells using *multi-laser flow cytometry*, which analyzes cells and particles as they flow past multiple lasers.

What did they find? Episodic exposures to $PM_{2.5}$ were associated with significantly elevated immune cells, including monocytes, natural killer cells, helper T-cells, and killer cells, but not B-cells. These findings provided evidence of nonspecific inflammatory immune responses. Additionally,

PM$_{2.5}$ exposures were associated with circulating endothelial microparticles, especially microparticles that indicated endothelial cell death.

The results showed the subclinical endothelial or vascular injury Aruni
and Arden had hypothesized. They wrote a draft manuscript presenting
these results and submitted it for peer review and publication in a scientific journal. One perceptive reviewer noted that if episodic exposures to
air pollution resulted in inflammatory immune cell responses and subclinical endothelial or vascular injury, there should be accompanying elevations in inflammatory *cytokines*. These cytokines are small proteins that
help regulate the body's inflammatory and immune responses. Could the
researchers look at these cytokines?

They could and they did. The team had retained frozen plasma from
all the blood draws. They sent the plasma samples to a biomedical laboratory that analyzed standard arrays of forty-two human cytokines and
growth factors using laser technology and two endothelial adhesion
markers. When Aruni and Arden analyzed the lab results for pollution-
related associations, they found that increased exposures to PM$_{2.5}$ air
pollution were associated with significant increases in crucial proinflammatory cytokines and endothelial adhesion markers. These results were
consistent with their hypothesis. The surprise came when they found
that several crucial growth factors (most notably, *epidermal growth factor,
EGF*) dropped with elevated levels of air pollution.

The team revised the original manuscript to add these new results.
The final version[60] reported evidence that episodic exposures to PM$_{2.5}$
air pollution were associated with three key biological responses: (1) elevated immune cells, indicating nonspecific inflammatory responses; (2)
elevated circulating endothelial microparticles, indicative of subclinical
endothelial injury; and (3) elevated proinflammatory cytokines but sup-
pressed growth factors, suggesting more vascular injury and less repair.
These findings were remarkably consistent with the general mechanistic
pathway of pollution-induced pulmonary and systemic oxidative stress
and inflammation outlined in figure 10.1.

As expected, none of the participants in the study experienced clinically significant symptoms or effects. That was good. A troubling aspect
of these findings, however, was that air pollution exposure contributed
to a slow, largely invisible, cumulative, and insidious process of cellular

damage—subclinical disease, ultimately advancing to clinical disease. This was true even in young, healthy adults. The study added to the growing body of scientific evidence that exposure to air pollution contributes to a cascade of inflammatory and immune responses. This cascade results in vascular injury, the initiation and progression of atherosclerosis, and related cardiovascular disease.

CONCLUSION

"How does air pollution cause health effects?"

The answer is still unclear. Scientists still don't fully understand the physiological or mechanistic pathways that link air pollution to cardiopulmonary disease. Given that complex interactive biological mechanisms are involved, a complete understanding may never be possible. But over the past few decades, there has been a dramatic increase in evidence of biological mechanisms and pathways linking air pollution exposure with cardiopulmonary diseases and even death.

Figure 10.1 necessarily provides only a stylized, incomplete, and selective illustration of biological mechanisms. It emphasizes generalized physiological pathways focused on the lungs, blood, blood vessels, heart, brain, and systemic inflammation and oxidative stress.

Other air pollution and health researchers have provided alternative illustrations of mechanistic pathways. For example, in 2010 the American Heart Association published an illustration that focused on three generalized pathways, including systemic oxidative stress and inflammation, particulate matter or constituents in the blood, and autonomic nervous system imbalance with subsequent specific biological responses.[61] Similar excellent illustrations of biological pathways, with different emphases, are available in the scientific literature.[62]

Although there is more to learn about the biological mechanisms of air pollution effects, key physiological pathways are increasingly being evaluated and understood. The observed health effects of air pollution are, indeed, biologically credible.

11

IS POLLUTION CONTROL WORTH THE ECONOMIC COST?

We've provided evidence in previous chapters that air pollution adversely impacts human health. Based on this evidence, can the overall burden of disease from these adverse health impacts be estimated? If so, can a dollar value of the overall cost of air pollution also be estimated and compared with the costs of reducing pollution? And, importantly, are there economic strategies or instruments that effectively and efficiently address these costs of air pollution?

In July 2012, the Swiss RE Center for Global Dialogue approached Doug and colleagues at the Harvard School of Public Health to discuss the effects of air pollution on populations worldwide, particularly in the economies of China, India, Brazil, and Mexico. Swiss RE is one of the world's leading providers of insurance, reinsurance, and other forms of insurance-based risk transfer.[1] Swiss RE was concerned about evidence that air pollution contributed to mortality and affected life expectancy. If so, air pollution would be a factor Swiss RE would have to assess in its business.

Doug and colleagues had informal discussions and colloquia with the Swiss RE group. Motivated by questions from these discussions, Doug and Arden conducted analyses that suggested multiple years of lost life expectancy in some of the most polluted cities throughout the world.[2] Such estimates of loss of life expectancy in highly polluted cities raise the question of the burden of disease attributable to air pollution worldwide, and ultimately the economic costs of these exposures.

BURDEN OF DISEASE

Burden-of-disease studies use exposure-response relationships estimated by the research on air pollution and health (discussed in previous chapters) and apply those estimates to populations worldwide. Table 11.1 provides estimates of global annual deaths from recent studies. One of the most important studies is the Global Burden of Disease (GBD) study conducted by the Institute for Health Metrics and Evaluation at the University of Washington.[3] This massive, ongoing collaborative study estimates attributable mortality and disability-adjusted life-years for eighty-seven risk factors in 204 countries.

Based on the GBD analysis, air pollution is the *single most significant environmental risk factor* contributing to the global disease burden. It is among the top five risk factors for global attributable deaths and disability-adjusted life-years. As reported in table 11.1, GBD estimates deaths attributed to outdoor (ambient) air pollution at approximately 4.5 million per year. Household air pollution (primarily from burning solid fuels for cooking and heating in the home) contributes to about 2.3

Table 11.1   Estimates of global annual deaths attributed to air pollution per year

Studies with key references	Pollution sources	Deaths (millions)
GBD 2019 Risk Factors Collaborators 2020	All sources	6.7
Health Effects Institute 2020	Ambient	4.5
Cohen et al. 2017	Household	2.3
Lelieveld et al. 2019	All sources Ambient Household	8.8 5.5 3.6
Burnett et al. 2018	Ambient	8.9
Vohra et al. 2021	Burning fossil fuels	8.7

Sources: See chapter notes 3 and 4.

million deaths. As reported in table 11.1, other studies have attributed even more global deaths to air pollution.[4]

LIFE EXPECTANCY REDUCTIONS

How does air pollution contribute to millions of deaths per year? As discussed in previous chapters, elevated exposures to air pollution increase the risks of acute respiratory and cardiovascular events, especially in those with existing cardiopulmonary disease.[5] Perhaps more disturbingly, air pollution contributes to excess annual deaths by accelerating chronic disease—essentially speeding the aging process[6] and reducing lifespan.

So, how *much* does air pollution reduce life expectancy, and how does that compare with other important risk factors? Michael Greenstone and Claire Qing Fan created an "Air Quality Life Index"[7] based on global air pollution estimates and the China Huai River study discussed in chapter 7.[8] Greenstone and Fan estimated that if the entire planet met the World Health Organization's guidelines for particulate matter air pollution, the average life expectancy for humanity would grow by 1.8 years. They also reported that the per capita life expectancy reduction caused by air pollution was comparable to cigarette smoking. Air pollution's contribution was substantially more significant than other well-known risk factors, such as alcohol and drug use, unsafe water or poor sanitation, road injuries, HIV/AIDS, malaria, tuberculosis, and conflict and terrorism.

But these results were based mainly on one remarkable, unique study from China. What is the evidence from the rest of the scientific literature?

Joshua Apte, an energetic civil and environmental engineering professor at the University of California, Berkeley, led a collaborative research effort that evaluated the impact of ambient $PM_{2.5}$ air pollution on global and regional life expectancy.[9] The team used data from the Global Burden of Disease and life-table methods to estimate reduced life expectancy for 185 countries. As expected, life expectancy varied across countries with different levels of air pollution. The average life expectancy reduction from ambient $PM_{2.5}$ was 1 year; from household air pollution, 0.7 year; and from all air pollution, 1.7 years.

For context, figure 11.1 presents Greenstone and Fan's estimates of life expectancy reduction for air pollution and the Apte team's estimates,

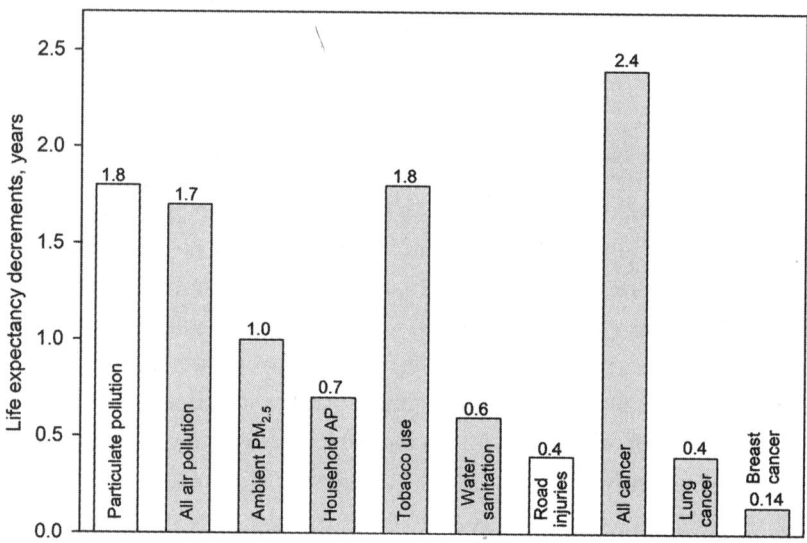

Figure 11.1

Estimates of life expectancy decrements for air pollution and selected other risk factors and causes of death. *Sources:* Modified from Michael Greenstone and Claire Quig Fan, "Introducing the Air Quality Life Index: Energy Policy Institute at the University of Chicago (EPIC)," November 2018 (white bars), and Joshua S. Apte, Michael Brauer, Aaron J. Cohen, Majid Ezzati, and C. Arden Pope, "Ambient $PM_{2.5}$ Reduces Global and Regional Life Expectancy," *Environmental Science & Technology Letters* 5, no. 9 (2018): 546–551 (gray bars).

along with life expectancy reductions associated with other risk factors and causes of death. Remarkably, although relative risks to an individual from smoking are much higher, average population-based life expectancy reductions from air pollution are comparable with cigarette smoking—mainly because of the widespread, ubiquitous exposure to air pollution. These results suggested that "the burden of disease from air pollution results in life expectancy decrements of a magnitude similar to those of other high-priority risk factors and diseases."

MONETIZED HEALTH COSTS OF AIR POLLUTION

In 2010, Arden was serving on the U.S. EPA Advisory Council on Clean Air Compliance Analysis. The Council was tasked with reviewing an EPA study of the benefits and costs of controlling air pollution.[10] On read-ing a review draft of the document, the approach and methods seemed reasonable. Much of it was based on epidemiological research that he, Doug, and other colleagues had helped conduct. The study was a good-faith effort to use existing evidence and to tally up and monetize the direct costs and benefits of controlling air pollution. The reading was a bit tedious. But then came the bottom-line estimates. The costs of public and private efforts to reduce air pollution by meeting Clean Air Act require-ments were estimated to total about $65 billion in 2020.

So, it seems efforts to clean up the air come with non-trivial costs. But what about the benefits? The benefit estimate was a whopping total of almost *$2 trillion*. Two trillion dollars in the United States alone seemed unbelievable—absurd. Furthermore, nearly all the benefits of controlling pollution came from the high value of reducing mortality.

Several efforts have been made to assess the global welfare costs of air pollution in financial terms. Table 11.2 provides recent monetary esti-mates of the global welfare costs of air pollution per year. Estimates come from Bjorn Larsen of the Copenhagen Consensus Center,[11] a joint study from the World Bank and the Institute for Health Metrics and Evalua-tion,[12] the Organisation for Economic Co-operation and Development,[13] the Lancet Commission on Pollution and Health,[14] and recent reports

Table 11.2 Estimates of global welfare costs of air pollution per year

Study	Pollution included	Year	$ (trillions)	Percentage of global GDP
Larsen 2014	AAP	2012	1.7 (US$)	2.5
World Bank and IHME 2016	AAP & HAP	2013	5.1 (PPP)	5.0
OECD 2016	AAP	2015	3.4 (PPP)	6.0
Landrigan et al. 2018	AAP & HAP	2015	3.8 (US$)	5.1
World Bank 2020	AAP	2016	5.7 (PPP)	4.8
World Bank 2022	AAP & HAP	2019	8.1 (PPP)	6.1

Notes: AAP = ambient (outdoor) air pollution; HAP = household air pollution; $ (PPP) = international dollars or purchasing power parity adjusted US$. *Sources:* This table is modified and updated from World Bank, "The Global Health Cost of PM2.5 Air Pollution: A Case for Action beyond 2021," International Development in Focus (Washington, DC: World Bank, 2022), doi:10.1596/978-1-4648-1816-5. Also see chapter notes 11–15.

from the World Bank.[15] The most notable finding in table 11.2 is that the costs of air pollution are massive—as much as *$8.1 trillion*, or approximately 5 or 6 percent of global gross domestic product (GDP).

These estimates of trillions of dollars of welfare costs from the health effects of air pollution, the equivalent of approximately 5 or 6 percent of global GDP, are enormous. How could such costs, in monetary terms, even be estimated?

For example, look at the 2022 World Bank estimates more closely.[16] World Bank researchers used the GBD 2019 estimates of premature mortality and morbidity attributable to $PM_{2.5}$ air pollution. The costs of premature deaths were monetized or given a dollar value by using estimates of the value of statistical life (VSL).

VSL is based on studies of how much people are willing to pay to reduce their risk of premature death. It represents how much a group or population is willing to pay to minimize the risk of death per person. *It is not the value of any individual or a reflection of society's assessment of human worth.* Rather, VSL provides a standard approach to evaluate human welfare losses associated with increased risks of death. Because VSL depends on income levels and various other socioeconomic factors, it differs across countries. For example, an average base VSL of $3.83 million is generated from studies of high-income member countries of the Organisation for Economic Co-operation and Development (OECD). In the World Bank study,[17] estimates of disease costs were based on years lived with disability (reflecting the duration and severity of disease), converted to days of illness, and monetized using weighted wage rates.

There is no perfect way to measure the human welfare costs of air pollution. Conscientious efforts by economists and health scientists to estimate these costs, however, indicate that air pollution contributes substantially to the burden of disease throughout the world. Humans generally place a high value on their lives and their health.

Again, it's important to note that the costs of air pollution health effects are not distributed evenly across the world. Figure 11.2 presents the annual welfare costs of $PM_{2.5}$ air pollution as a percentage of GDP globally and by regions of the world. The relative welfare costs of air pollution are much higher in South and East Asia, with household air pollution (HAP) contributing substantially to the health costs. In North

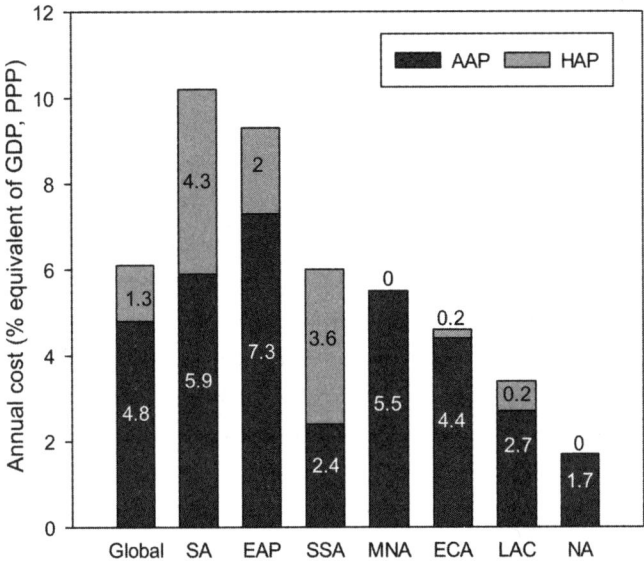

Figure 11.2

Estimated annual welfare cost from adverse health effects of $PM_{2.5}$ air pollution in 2019 globally and by region, percentage equivalent of GDP. AAP = ambient air pollution; EAP = East Asia and Pacific; ECA = Europe and Central Asia; HAP = household air pollution; LAC = Latin America and Caribbean; MNA = Middle East and North Africa; NA = North America; SA = South Asia; SSA = sub-Saharan Africa. *Source:* Modified from World Bank, "The Global Health Cost of $PM_{2.5}$ Air Pollution: A Case for Action beyond 2021," International Development in Focus (Washington, DC: World Bank, 2022), doi:10.1596/978-1-4648-1816-5.

America (the United States and Canada), the relative welfare costs of pollution are much lower (about 1.7 percent of GDP).

ECONOMIC BENEFITS OF CLEAN AIR

Why are health-related welfare costs of air pollution so much lower in the United States and Canada? One apparent reason is the significant disparities in household pollution internationally. In the United States and Canada, exposures from indoor solid-fuel use are tiny, contributing minimally to morbidity and mortality (see figure 11.2). The latest GBD report indicates that household air pollution declined more than most risks from 2010 to 2019 and that this decline was linked to social and economic development.[18] But, as we can see in figure 11.2, household air pollution continues to impose substantial health welfare costs in South and East Asia and sub-Saharan Africa. The Word Bank recommends public policy and economic efforts to foster the transition to cleaner fuel for cooking, space heating, and other domestic and household uses.[19]

Another reason the welfare costs of air pollution are much lower in the United States and Canada is the lower levels of outdoor or ambient air pollution compared with polluted parts of the world. Lower pollution in the United States is the result of concerted public policy efforts to improve and protect air quality. The U.S. Clean Air Act and amendments since 1970 have had their effect, reducing air pollution and related health and environmental damage.

In 1997, 1999, and 2011, the U.S. EPA released a series of studies that evaluated the benefits and costs of the Clean Air Act.[20] The primary economic benefits of the Clean Air Act and other clean air efforts are the health costs avoided by protecting air quality. Evidence shows that the health benefits of clean air policies are massive and much larger than the costs associated with implementing and conducting clean air policies.

The U.S. Office of Management and Budget conducted another notable and relevant analysis. It concluded that control of $PM_{2.5}$ air pollution provided more public benefit per cost than any other federal regulation or unfunded mandate.[21]

It is difficult to disentangle and quantify all the benefits and costs of air pollution. However, based on over fifty years of concerted efforts to reduce air pollution in the United States, is there evidence that controlling air pollution, in general, harms the economy?

Figure 11.3 compares changes in economic growth and activity with air pollution emissions. This graph begins in 1970, the year of the Clean Air Act, the establishment of National Ambient Air Quality Standards, and the establishment of the U.S. Environmental Protection Agency.[22] The graph shows that air pollution emissions declined substantially—an approximately *80 percent* decrease over fifty years. However, the population and vehicle miles traveled grew markedly. Economic activity, as measured by GDP, grew by nearly 300 percent. Based on this evidence,

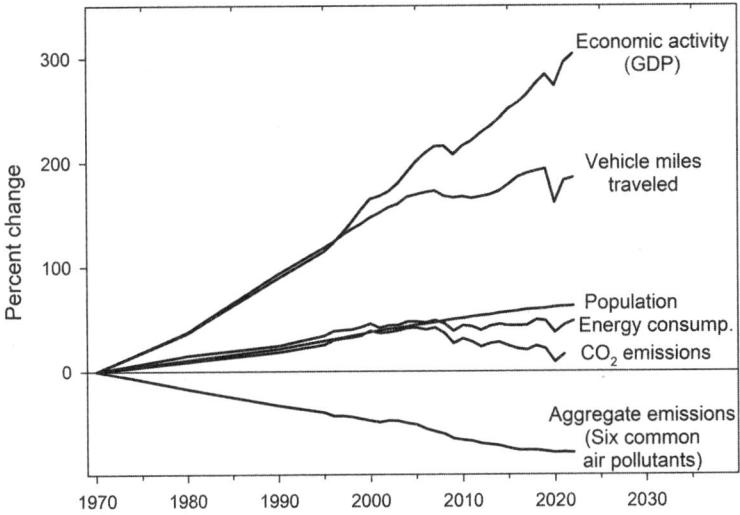

Figure 11.3

Comparison of changes in economic activity with changes in air pollution emissions from 1970 to 2022. *Source:* Plot modified using data from U.S. EPA, "Our Nation's Air: Trends through 2022," U.S. Environmental Protection Agency, 2023.

reducing and controlling air pollution emissions is consistent with— maybe complementary to—a dynamic, growing, modern economy.

What would economic growth in the United States (as represented in figure 11.3) have been *without* successful efforts to reduce and control air pollution? It is uncertain, but the absence of clean air efforts would have resulted in more polluted air, less visibility, and more respiratory, cardiovascular, and lung cancer disease and death.

OTHER ECONOMIC BENEFITS OF CLEAN AIR

This book focuses on the human health costs of air pollution. However, health cost savings are not the only benefits of cleaner air. Air pollution affects ecological health, natural resource productivity, and aesthetic values, directly impacting human welfare. For example, as noted in chapter 1 (and illustrated in figure 1.3), many related and common air pollutants come from the same combustion sources and contribute to health effects and effects on the climate system. Air pollution impacts forest and agricultural plant growth and productivity. It also contributes to the damage and soiling of buildings, sculptures, and other materials. Another obvious human welfare cost of fine particulate matter air pollution is its negative impact on visibility. Reduced visibility due to air pollution impacts recreational and residential values and decreases human welfare and quality of life. Estimates in the United States indicate that the costs of air pollution's effects on agricultural and forest productivity, materials damage, and visibility are substantial, but much smaller than the health costs.[23]

ECONOMIC INSTRUMENTS OR APPROACHES TO ADDRESS AIR POLLUTION COSTS

The evidence shows that reducing and controlling air pollution have enormous positive effects on human health and welfare. The obvious question then becomes: What can we do about it?

We can and should do much individually to reduce air pollution. But the solutions are not simple. Adequately addressing these issues requires appreciating that air pollution is a large, systemic economic problem. As a matter of public policy, and within the framework of national and global economies, are there ways to reap the benefits of cleaner air? A full discussion of the details of public policy approaches and economic instruments that can help address air pollution is beyond the scope of this book. But basic knowledge of the range of these approaches is important in searching for working solutions to the problem of pollution.

The following are some possible approaches.

1. *Do nothing.* In the spirit of the doctrine of laissez-faire (see chapter 2), one public policy approach to air pollution is to largely ignore its costs and losses to society. This approach appeals to those who adhere to libertarian economic and political philosophies. Under that ideology, free markets and minimal government intervention will somehow address the air pollution problem.

But there are evident problems with this do-nothing or "let it be" approach. The benefits of polluting activities are largely *internalized* (reaped by the polluter), creating incentives to increase polluting activities. The costs of air pollution are largely *externalized* (incurred by others in society), giving polluters little self-interest in reducing or controlling pollution. Because externalized costs are not folded into production costs, the market prices of products from polluting activities are lower than their true total costs.

If the resulting externalized costs and deadweight loss to society were minimal, this "laissez-faire" approach might not matter much. But those costs are high, conspicuous, damaging to human health and welfare, and detrimental to the environment and climate. Considering this evidence, doing nothing and ignoring the massive, externalized costs of pollution seems reckless from a public policy perspective, self-serving on the part of polluters, and harmful to societal economic and public health interests.

2. *Moral suasion.* A second public policy approach to dealing with the externalized costs of air pollution would be moral suasion, "jawboning," or appeals to ethical or responsible behavior. This approach could include efforts to educate polluters about pollution costs and appeal to their sense of civic, social, and moral duty. A well-known example of this approach was the U.S. Forest Service's creation of Woodsy Owl and the motto, "Give a hoot—don't pollute," or a more recent slogan, "Lend a hand—care for the land." Woodsy Owl, created in 1971, was featured in various educational films and moral suasion efforts.[24]

"Give a hoot—don't pollute" campaigns may have effectively persuaded people not to litter on hiking trails or in public parks. But how effective or practical are moral suasion approaches with major air pollution sources? Many of these sources blatantly deny their contribution to pollution or adverse public health effects due to it. Such educational and moral suasion approaches have little chance of success when shrewdly countered by well-funded "merchants of doubt" who have sophisticated public relations and political lobbying campaigns. Industry campaigns to obscure the truth and cast doubt on the enormous, externalized costs imposed upon society have been well documented elsewhere.[25] More effective and compelling public policy action is required.

3. *Regulation.* Traditional *regulatory* actions are common, well-known public policy approaches addressing air pollution. Government regulation or command-and-control regulatory approaches include setting

and enforcing ambient air quality or emission standards for automobiles, trucks, factories, power plants, and other pollution sources. These approaches could include establishing and enforcing clean air "best practices" or "standards of performance" for multiple polluting industries and activities. They could include mandated scrubbers on smokestacks, vehicle emissions inspections and maintenance programs, bans on open pit burning and the burning of solid fuels for space heating, and much more. Regulatory actions could be strictly health-based, determined by benefit-cost analyses, or based on other criteria. As we've noted, with the establishment of the Clean Air Act, National Ambient Air Quality Standards, and a myriad of regulatory requirements, the United States has had success in reducing and controlling air pollution.

The regulatory approach, however, has its own deficiencies and limitations. An ideal regulatory approach assumes regulatory management that is wise and competent, and regulations that are acceptable, well-enforced, efficient, low-cost, and politically stable. In reality, government regulation relies on a "command-and-control" approach that is likely inefficient, difficult to enforce, and could result in an overreaching, growing, and more intrusive government. The overall regulatory effort can become extensive, complex, and politically and bureaucratically unwieldy.

In addition, polluters sometimes avoid regulations or favorably alter them with "rent-seeking" or self-interested political behavior. Polluters seek to manipulate regulatory efforts to obtain economic benefits while passing off costs to others. Much of the controversy and debate over the health effects of air pollution are proxy battles to avoid regulatory action. Therefore, more efficient, cost-effective, and less complex approaches that are less susceptible to manipulation would be helpful.

4. Cap and trade. Another approach to controlling air pollution includes emissions trading schemes, often called cap and trade. In a cap-and-trade or emissions trading system, the government limits—or caps—overall air pollution emissions and then establishes transferable permits (or allowances) to emit pollutants.

A substantial appeal of cap and trade is that it is a market-based approach that uses the power and ingenuity of competitive markets to find efficient, cost-effective, and acceptable techniques and processes to control pollution. The sum of pollution permits, or allowances, is capped at a constrained pollution level. Permits can be bought or sold, and the market will determine the price for permits. In effect, this approach requires polluters to incur at least part of the cost of pollution by requiring them to have permits to pollute. Cap and trade incentivizes emissions reductions by allocating pollution permits and allowing unused permits to be sold at the market price. Low-cost pollution controls could allow some companies to profit from cap and trade.

The cap-and-trade or emissions trading approach, however, has some substantial challenges. For example, it is difficult to determine the ideal or optimal cap for overall emissions. And it isn't easy to allocate permits or to monitor and enforce their trading. The cap-and-trade approach further assumes that the market for pollution permits is feasible, enforceable, and politically acceptable.[26] Is there a more straightforward approach?

5. *Emissions taxes.* Another approach to controlling air pollution is emissions or corrective taxes. The approach is conceptually simple— the government places a price or tax on pollution emissions. Ideally, this corrective tax would be calibrated to equal the marginal costs of pollution to society. This marginal cost of pollution would be incurred by polluters themselves—providing appropriate incentives to control pollution. If the price of reducing pollution is less than the emissions tax, the polluter finds it financially advantageous to reduce emissions.

The appeal of emissions taxes compared with the cap-and-trade approach is its relative simplicity. Just set the tax equal to the marginal social cost of pollution. Unfortunately, it is not always clear exactly what the marginal cost is.

Emissions taxes have another appealing aspect. They provide incentives to curb air pollution. Some taxes have perverse incentives and are inefficient. For example, income taxes can result in incentives to work less and be less productive. Well-designed emissions taxes, however, are corrective by providing incentives to reduce pollution. Further, revenues from emissions taxes could be used to offset other less-efficient taxes, offer tax rebates, or provide funds for other public policy efforts to mitigate pollution costs.

Emissions taxes and cap-and-trade schemes are conceptually and theoretically similar. Both are *market-based* approaches. Unlike a regulatory approach, with market-based approaches, the government doesn't determine *how* polluters should clean up. These approaches are designed to internalize the costs of air pollution while allowing economic market incentives, innovation, and flexibility.

EXAMPLE OF A MARKET-BASED APPROACH

The Climate Leadership Council proposed a market-based approach to address air pollution from fossil fuels. This proposal originated with prominent Republican leaders and economists in the United States. It garnered bipartisan support, corporate support, and the most prominent public statement of support from economists in history. Signatories include over 3,500 U.S. economists, twenty-eight Nobel Laureate economists, fifteen former chairs of the Council of Economic Advisers, and four former chairs of the Federal Reserve. The proposal is billed as a "carbon dividends plan" and addresses concerns about fossil fuel carbon emissions and climate impacts. It is a well-thought-out and well-promoted emissions pricing or corrective tax plan.

The four basic tenets of the proposed carbon dividends plan are: (1) Impose a fee or tax for carbon emissions; (2) return to U.S. families all net revenues from these taxes as "carbon dividends" or lump-sum rebates; (3) reduce redundant carbon-related regulations that are less efficient, allowing businesses to focus on innovative fuels and technology; and (4) impose a border carbon adjustment on foreign goods.

Even this highly ambitious market-based carbon emissions tax plan would not eliminate all air pollution regulations. It would only replace redundant rules governing carbon emissions. Although it would likely reduce other fossil fuel emissions, it would not address non-carbon emissions directly. As such, this plan would complement various regulatory efforts to reduce air pollution in general.

Supporters of the carbon dividends plan estimate that this approach would dramatically reduce pollution *and* pay for itself because it is a more efficient corrective tax. Equal lump-sum dividends of rebates would provide fairness and political viability. Most Americans, especially the most vulnerable, would receive more in dividends or lump-sum rebates than they would pay in increased energy prices. Further, the plan would promote economic growth and innovation.[27]

CONCLUSION

"Is pollution control worth the economic cost?"

The answer is reasonably evident. Although the health costs of air pollution are enormous—much more extensive than Doug and Arden imagined when they first started studying the health effects of air pollution—the benefits of clean air far outweigh the costs.

Air pollution is one of the largest risk factors contributing to the global burden of disease. Air pollution contributes to millions of excess deaths per year and substantially reduces life expectancy. Global human welfare costs per year are as high as US$8.1 trillion and approximately 5–6 percent of global GDP. By almost any relevant measure, the health costs of air pollution are *not* small.

But does controlling air pollution harm our economy? No, not generally. There are various public policy approaches to reducing and controlling air pollution, including market-based efforts. Ambient air quality standards and related regulatory actions have already had remarkable success in the United States and elsewhere. The expanded use of complementary market-based strategies may provide more efficient and cost-effective approaches to reducing and controlling air pollution.

Empirical evidence over the last fifty years is relatively compelling on one critical issue: *controlling air pollution emissions and protecting air quality is consistent with and complementary to dynamic, growing, modern economies.* Further, humans generally place a high value on their lives and their health. Therefore, air quality is among society's economic choices. Clean air is an economic good that contributes to human well-being, human capital, and favorable environmental amenities. Reducing pollution and protecting clean air contribute to economic prosperity, human well-being, and improved public health. Clean air is integral to public health, social welfare, and a robust economy.

12

WHEN WILL THE EVIDENCE END THE CONTROVERSY?

"Modern air is a little too clean for optimum health."

Robert Phalen made this provocative statement.[1] Some used it to disparage him when he was appointed to serve as a member of the EPA's Clean Air Scientific Advisory Committee.[2] This and similar statements were also used to label Phalen an "air pollution denier." He has undoubtedly contributed to the scientific and public policy controversies regarding air pollution and health.[3]

Phalen, however, is no charlatan. He is a well-trained inhalation toxicologist, a professor of medicine, director of the Air Pollution Health Effects Laboratory Center at the University of California, Irvine, and a member of the American Association for the Advancement of Science. In 2002, he published an insightful and useful book, *The Particulate Air Pollution Controversy*.[4] Yet Phalen seems to revel in perpetuating controversy about the health effects of air pollution. He has asserted that his "most important role in science is causing trouble and controversy."[5]

For Arden and Doug, Robert Phalen is a bit of an enigma. Their goal as researchers on the health effects of air pollution is not to cause trouble but to understand and find ways to alleviate the contribution of air pollution to disease and death. They conduct research not to provoke controversy but to provide evidence, information, and scientific consensus that helps resolve disputes. Although they recognize that legitimate debate can stimulate research and is a necessary part of the scientific process, controversies for the sake of controversies are rarely productive.

The topic of controversy invokes the question asked in this final chapter: When will the evidence end the controversy? We'll address this question as well as a further question: When will scientific evidence be so compelling that public policy efforts for clean air are no longer controversial?

Scientific research regarding air pollution and human health has advanced dramatically over the last several decades. This story includes the historic "killer smog" episodes of the 1930s–1950s, with air so thick and polluted that visibility dramatically deteriorated. There were obvious increases in respiratory and cardiovascular disease and death. The story includes public policy efforts to reduce air pollution, informed by scientific evidence. For example, in 2021 the World Health Organization (WHO) proposed a stringent air quality guideline for annual concentrations of $PM_{2.5}$ of less than 5 µg/m^3. This level would require substantially reduced air pollution in almost all major populated areas.[6] This story also includes a long-running, often rancorous, scientific and public policy debate over reducing and controlling air pollution.

SUMMARIZING THE SCIENTIFIC AND PUBLIC POLICY DEBATE

We can summarize the public policy controversy about air pollution as a back-and-forth debate of various assertions and responses. In this stylized debate, the *assertions* represent interests that resist measures to reduce and control air pollution. These assertions also reflect a general skepticism about the scientific evidence for adverse health effects caused by air pollution. The *responses* represent research efforts and reflect the evidence that air pollution contributes to human morbidity and mortality.

Assertion 1 Economic enterprise and markets should be as free as possible with minimal restrictions (see the doctrine of laissez-faire). If pollution occurs, it is an unavoidable consequence of prosperity. Pollution "smells like money."

Response Economic activity, free enterprise, industrial production, and technological progress all can and do make major contributions to human health and welfare. The need for these activities is not in dispute. Some activities, however, are also substantial pollution producers. If pollution is unabated, it can result in unsightly, smelly, noxious, unhealthy pollution. Extreme air pollution episodes in the Meuse Valley, Donora, and London produced compelling evidence that air pollution contributes to disease and death, at least at extremely high levels. This evidence motivated public policy efforts to control air pollution in the United Kingdom and the United States (see chapter 2).

Assertion 2 There may have been some adverse health effects at extremely high levels of air pollution, but these episodes were rare and uncontrollable weather phenomena. Most levels of air pollution are safe. Common, low to moderate levels of pollution are not harmful.

Response Studies show that even moderate levels of air pollution are harmful. For example, the intermittent operation of a steel mill in a populated mountain valley demonstrated that moderate levels of air pollution had a harmful effect on children's respiratory health. Further, panel and daily time-series studies of mortality and hospitalizations established that short-term, day-to-day exposures to common pollution levels are associated with adverse health effects. These effects include increased cardiovascular and respiratory disease and increased risk of death (see chapter 3).

Assertion 3 Results from these studies may have been confounded by increased viral infections that coincidently occurred with the steel mill operation. Associations from the time-series studies may have resulted from inadequate control for weather or some other potential confounders. Therefore, we need these studies replicated and validated in many areas and with improved methods.

Response The early daily times-series studies *were* replicated and validated. Researchers used alternative approaches to control for weather, and the results were unchanged. Daily time-series studies were conducted in hundreds of other cities in many different settings by many researchers with similar results. To answer any lingering doubts, a new analytic design, the case-crossover approach, was developed. This approach controls for subject-specific differences, seasonality, time trends, and day of week by design rather than statistical control. This approach was applied to death, acute cardiovascular events (heart attacks and strokes), and acute respiratory events. Again, similar results were obtained (see chapter 3).

Assertion 4   Short-term exposures have only minor, short-lived effects. Only the old and frail are affected by short-term air pollution exposures and would have died soon anyway. Long-term exposures are most relevant to health. Therefore, we need studies that evaluate long-term air pollution exposures in terms of pertinent health measures.

Response   More than the old and frail were affected by short-term air pollution. Yet it's true that long-term exposures have much more significant effects on health than short-term exposures. Population-based cross-sectional studies demonstrate that differences in long-term average air pollution in metro areas are associated with differences in mortality rates, especially for respiratory and cardiovascular mortality and lung cancer. These associations are observed even when controlling for socioeconomic factors, smoking prevalence, and other variables (see chapter 4).

Assertion 5   Studies of population-based data can't be trusted. It is impossible to control everything in a population. What about the ecological or population fallacy? Impacts on individuals cannot be inferred from effects on populations. Studies based on individual data are needed. They may be expensive and take a long time to complete, but we need prospective cohort studies.

Response   An initial prospective cohort study that took about twenty years to complete found evidence that mortality risks—especially cardiopulmonary mortality—were strongly associated with air pollution. Motivated by this study, a parallel, complementary, and much larger cohort was analyzed and reported. The analysis found similar results (see chapter 4).

Assertion 6   We can't fully trust the results from these two early cohort studies. The estimated effects of air pollution are surprisingly large, much larger than observed from the daily time-series and case-crossover studies. The initial two studies must be validated with independent auditing and analysis.

Response An independent research team received access to all the critical data from the initial two cohort studies. That team audited the data and found they were of high quality. The team validated and replicated the originally reported results. Sensitivity analyses and sophisticated modeling indicated that the originally reported pollution-mortality associations were remarkably robust.

Other teams conducted extended analyses of both cohorts using longer follow-ups, more advanced statistical modeling, improved air pollution exposure assessment, and other refinements. All these studies showed that long-term exposures to $PM_{2.5}$ air pollution were associated with mortality risk, especially cardiopulmonary and lung cancer mortality (see chapter 4).

Assertion 7 The initial cohort studies are suggestive, but not conclusive. Other studies are needed using cohorts from other areas of the world.

Response Studies of other cohorts were conducted, including studies in the United States, Canada, Europe, and China. Meta-analyses of these cohorts provide further analytic evidence that long-term exposure to $PM_{2.5}$ air pollution contributes to lung cancer, cardiopulmonary disease, and death (see chapter 5). But there's more. The health burden of air pollution is not just borne by the old and frail. All ages bear it. The evidence suggests that air pollution begins to contribute to disease early in life. Further, it is borne disproportionately by vulnerable groups: racial and ethnic minorities, people of low socioeconomic status, and those constrained to living in highly polluted communities and highly polluted parts of the world (chapter 6).

Assertion 8 All these cohort studies have a problem. They are based on individual data that are not publicly available. They are "secret science" and should not be trusted or used to inform public policy unless their data are publicly available.

Response Demanding the release of personal health information ignores legitimate ethical and legal constraints established to protect research subjects. Confidentiality agreements and ethical and legal restrictions limit the ability to release private and confidential study-participant data publicly.

The results of these studies were peer-reviewed and published in open scientific literature, and some have been audited and reanalyzed by independent research teams. Further, some studies have used cohorts constructed from public data (see chapter 5) precisely to address this issue. To reject legitimate and crucial scientific evidence from cohort studies that use individual data (as well as other studies that use restricted clinical data) because confidential data are protected would be a travesty.

Assertion 9   These studies do not necessarily demonstrate causal associations. Correlation is not causation. There is a need to demonstrate that reductions in air pollution result in reductions in adverse health effects. Further, there is a need for studies that formally and methodologically address issues of causal inference.

Response   Multiple studies demonstrate that reductions in air pollution are linked with improvements in health (see chapter 7). Furthermore, a large amount of epidemiological research has used remarkable natural experiments, quasi-experimental designs, and causal modeling approaches to rigorously evaluate and explore causal inference (see chapter 9).

Assertion 10   Even with all the epidemiological evidence, what is needed is an understanding of the biological mechanisms that explain why breathing polluted air adversely affects respiratory and cardiovascular health.

Response   There is still much to learn about biological mechanisms. But research has already demonstrated crucial underlying physiological pathways through which air pollution harms the body at a cellular level. There is ample evidence that the relationship between air pollution and respiratory and cardiovascular disease is biologically credible (see chapter 10).

Assertion 11   Even if there are some health costs of air pollution, these costs are small, and efforts to control pollution could harm our economy.

Response   The costs of air pollution are *not* small. Air pollution contributes substantially to the global burden of disease, decreased life expectancy, and human welfare losses. Clean air, clean water, and wholesome food are integral to public health, social welfare, and economic prosperity (see chapter 11).

GROWING CONSENSUS OVER SCIENTIFIC EVIDENCE

As illustrated in this stylized scientific and public policy debate, some air pollution controversies and disputes have been partially resolved. Over time, the accumulating and increasingly compelling scientific evidence regarding the health effects of air pollution has reduced controversy over the scientific evidence. There is a growing consensus that air pollution contributes to various diseases, including respiratory, cardiovascular, and cancer (especially lung cancer).

This consensus is reflected in multiple authoritative reports and reviews. The following are a few of them:

- The American Lung Association's 2022 State of the Air report succinctly states: "Years of scientific research have clearly established that particulate pollution and ozone are a threat to human health at every stage of life, increasing the risk of premature birth, causing or worsening lung and heart disease, and shortening lives."[7]
- The European Society of Cardiology (ESC) Working Group on Thrombosis, the European Association for Cardiovascular Prevention and Rehabilitation, and the ESC Heart Failure Association issued a position paper in 2022. It stated: "There is now abundant evidence that air pollution contributes to the risk of cardiovascular disease and associated mortality, underpinned by credible evidence of multiple mechanisms that may drive these associations. In light of this evidence, efforts to reduce exposure to air pollution should urgently be intensified and supported by appropriate and effective legislation."[8]
- The American Heart Association (AHA) issued a statement: "Ambient air pollution is the world's leading environmental risk factor, contributing more to global morbidity and mortality than many other common risk factors. . . . Because of an aging population and growing number of at-risk individuals (e.g., patients with cardiovascular disease), significant reductions in $PM_{2.5}$ levels will be needed to prevent an increase from this already unacceptably high public health toll."[9]

- The *Journal of the American College of Cardiology* published a state-of-the-art review that indicated: "Fine particulate matter air pollution < 2.5 μm ($PM_{2.5}$) is the most important environmental risk factor contributing to global cardiovascular (CV) mortality and disability. . . . As such, ambient $PM_{2.5}$ poses a major threat to global public health."[10]
- The International Agency for Research on Cancer (IARC) determined in 2013 that particulate matter air pollution was a human carcinogen for lung cancer.[11] A more recent overview of the evidence of air pollution and cancer concluded, "Outdoor air pollution poses an urgent worldwide public health challenge because it is ubiquitous and has numerous serious adverse human health effects, including cancer. Currently, there is substantial evidence from studies of humans and experimental animals as well as mechanistic evidence to support a causal link between outdoor (ambient) air pollution and especially particulate matter (PM) in outdoor air, with lung cancer incidence and mortality."[12]
- The national academies of five countries, including the U.S. National Academy of Sciences, coauthored an expert consensus document on air pollution and health science. The document states, "Poor air quality threatens human life, population health, and the future prosperity of children. Air pollution also threatens the sustainability of the earth's environment, as clean air is as vital to life on earth as clean water. The scientific evidence is unequivocal: air pollution can harm health across the entire lifespan. It causes disease, disability, and death, and impairs everyone's quality of life."[13]

Many other authoritative reports and reviews have reached similar conclusions. These include the U.S. Environmental Protection Agency's Integrated Science Assessments,[14] the Health Effects Institute's *State of Global Air*,[15] and the World Health Organization's *Global Health Guidelines*.[16]

CONCLUSION

When will we reach a point where the scientific evidence is so compelling that public policy efforts on behalf of clean air are no longer controversial?

We should be at that point now, but we are not. As illustrated in the stylized debate presented above, as some issues are resolved, alternative objections to reducing and controlling air pollution are generated to replace them. In some ways, air pollution and health research intended to provide scientific resolution instead perpetuates and expands the controversy. Furthermore, as research questions about the health effects of air pollution become resolved, the debate over clean air efforts necessarily and appropriately shifts to public policy options and economic, political, and related issues.

There are several reasons that there will likely be continued controversy and debate over public policy efforts to control air pollution.

1. Externalized costs. The costs of air pollution are classic negative externalities. Polluters and polluting activities do not pay most of the costs of air pollution. As discussed in chapter 2, air pollution represents a cost the polluter is not required to pay. It is easier to make money if they can keep the revenues (internalize the benefits) and pass off some costs (externalize the costs).

Because of the profit motive, polluters may always try to avoid the costs of their pollution. And as they become aware of the imposed health costs, breathers will try to implement public policy to reduce these costs. Conceptually, economists understand that externalized costs can result in substantial market failures.[17]

As discussed in chapter 11, market-based approaches can help address these market failures. However, an essential aspect of the public policy debate over air pollution has less to do with the degree of certainty regarding air pollution and health science and more with the distribution of costs and benefits of pollution and polluting activities.

2. Health costs are not directly or easily observed. During the time this chapter was being written, a dramatic event occurred at a rural train crossing in Missouri. An Amtrak passenger train traveling at high speed struck a loaded dump truck at a railroad crossing. The truck driver and three train passengers died; about 150 people were injured.[18] Unlike the obvious tragedy of this train wreck, death and injury from air pollution to individuals and the public are not as easily or directly identified. Often, the health costs of air pollution are hidden or not fully understood. These costs include excess disease and death distributed among the people who breathe polluted air. Air pollution is, essentially, a slow emergency, and our brains aren't wired to respond.

When air pollution contributes to disease or death, pollution is rarely mentioned on medical records or death certificates. The relatively obvious case of Ella Kissi-Debrah, one of the children profiled in chapter 1, was a rare exception. The slow, cumulative, insidious contributions of long-term exposure to air pollution to chronic respiratory, cardiovascular, or cancer diseases are not easily recognized in our daily lives.

Pollution's contribution to disease often mimics the body's natural aging process. Even when elevated air pollution triggers acute cardiopulmonary events (such as heart attacks, strokes, or acute respiratory infections), it is difficult to quantify the relative contribution of air pollution versus other risk factors.

3. Efforts to obfuscate. Even when the scientific evidence of air pollution's health effects is compelling, some people are not satisfied by any amount of statistical, epidemiological, or even physiological evidence. Some perpetually argue that science is incomplete and uncertain. They imply that public policy action should wait until we know all the details with certainty. This "paralysis by analysis" is often a concerted strategy.[19] Others argue that the scientific community consists of purveyors of corrupt, junk, or "secret science." They say that air pollution and health researchers are motivated by self-interest and political ideology and cannot be trusted.[20]

Such efforts to obfuscate have been used before. For example, there is an enormous amount of scientific evidence of the pernicious health effects of cigarette smoking. Yet even with this incontrovertible evidence, there continues to be controversy regarding the appropriate public health policies related to cigarette smoking.[21] People still smoke. And some continue to support ongoing efforts to mass-market cigarettes.

4. When is the air clean enough? It depends on what we mean by "clean enough." From a medical or public health perspective, optimal levels of air pollution would be levels where there are no longer adverse health effects or risks. From an economic or public policy perspective, the optimal level of air pollution is where the incremental cost of reducing pollution exceeds the benefits of reducing pollution.[22]

As progress is made to reduce pollution, there will be fewer health costs. However, as the air becomes cleaner and cleaner, it will be increasingly challenging to continue to *reduce* pollution, and so the incremental costs of additional pollution reduction will rise. At some point, the benefits of pollution reduction will drop below the pollution abatement costs.

Concerted and successful efforts to control air pollution will ultimately lead to debates over when the air is clean enough. These are and will be legitimate and ongoing debates. At some point, resources used to reduce air pollution might be more effectively used in other ways to improve health and welfare.

CLIMATE CHANGE, AIR POLLUTION, AND HEALTH

Speaking of slow emergencies, there are complex but inextricable links between air pollution, human health, and climate change.[23] A full discussion of these links goes well beyond this book. But burning fossil fuels contributes to air pollution that adversely affects human health and impacts the global climate. Efforts to reduce fossil fuel–related pollutants that influence the global climate also reduce co-pollutants that directly impact human health (see figure 1.3).

For example, in 2015, the U.S. EPA announced a Clean Power Plan to reduce carbon pollution from power plants. The plan's primary objective was to address climate change issues. It would also substantially reduce pollutants contributing to adverse health effects, including fine particulate matter air pollution ($PM_{2.5}$). It was estimated that reducing health-related pollutants would avoid thousands of premature deaths, heart attacks, and asthma attacks, and hundreds of thousands of missed work and school days annually.[24]

Should these co-benefits be included in evaluations of the benefits and costs of the plan? There has been substantial debate about using health co-benefits to justify policy efforts to address climate change.[25] Some are concerned that climate change is too crucial to muddy the water with other issues. Yet adequate and rigorous accounting for co-benefits (and

costs) allows for evaluating efficient, cost-effective approaches to generate maximal climate and public health benefits.[26]

Another challenge related to climate protection efforts by reducing fossil fuel–related pollution is that most climate benefits occur many years
after the reduction in emissions. Future benefits are generally discounted
or valued much less than current benefits. But reductions in fossil fuel
emissions can have immediate and direct benefits to human health. Less
immediate but substantial benefits to human health come from protecting the global climate.

The health benefits of cleaner air can become embroiled in conflated,
rancorous debates over efforts to reduce the fossil fuel emissions that
impact health and global warming. In *They Knew: The U.S. Federal Government's Fifty-Year Role in Causing the Climate Crisis*, James Speth provides
a fascinating, yet troubling, account of the controversies and debates
related to fossil fuel use, the climate crisis, and the U.S. government's role
in this crisis.[27]

There is an important aspect of the climate debate that is highly relevant to the subject of this book. Efforts to reduce greenhouse gas emissions generally result in reductions of co-pollutants (including $PM_{2.5}$),
providing direct and immediate benefits to human health. The mutual
benefits from efforts to reduce co-pollutants that affect human health
and contribute to global warming should provide additional and more
urgent impetus to reduce and control fossil fuel–related air pollution.
Yet the inextricable and complex links between air pollutants, human
health, and climate change complicate the controversy over public policy
efforts to control pollution.

ARE WE THERE YET?

Over the last several decades, there has been enormous progress in understanding the health effects of air pollution. Exposure to air pollution is
ubiquitous and involuntary. It contributes to the risk of disease and death
and is responsible for substantial losses in social welfare and well-being.
Air pollution is a significant contributor to the burden of disease globally.
From one perspective, these findings are disturbing, discouraging, and
dismal news.

From another perspective, the findings about the health effects of air pollution are good news.[28] Yes, air pollution significantly contributes to various diseases, including respiratory, cardiovascular, lung cancer disease, and more. However, the evidence provided in this book demonstrates that some of the health effects of air pollution are reversible. And exposure to air pollution can be reduced and controlled—resulting in substantial disease prevention and considerable benefits to human health and welfare. Reducing air pollution improves health, visibility, and the environment.

Well-designed efforts to reduce fossil fuel emissions and related pollutants that impact human health can be part of a strategy that helps reduce co-pollutants contributing to global warming and ocean acidification. The potential human health and environmental benefits are massive.

As Austin Bradford Hill presciently reminded us more than fifty years ago: "All scientific work is incomplete—whether it be observational or experimental. All scientific work is liable to be upset or modified by advancing knowledge. That does not confer upon us a freedom to ignore the knowledge we already have or to postpone the actions that it appears to demand at a given time."[29]

The stakes are high. But the work is clearly before those who want clean, healthy air for all who breathe.

ACKNOWLEDGMENTS

We are extremely grateful to Ronda and Jeanne, our fantastic wives who supported not just our efforts to write this book but our decades of research and related academic scholarship. In so many ways, they are our best and most important collaborators.

We thank Kara West for her early arguments that this book tells an important story and for her willing and skilled help with editing the book from start to finish. We are grateful to Anne Connor and Regina Higgins, accomplished writers and editors who provided insightful developmental editing. They substantially helped improve the quality and readability of the book.

We especially thank Beth Clevenger, with the MIT Press, who was responsive and encouraging and who provided forthright, constructive, and perceptive guidance. Anthony Zannino, Deborah Cantor-Adams, and Stephanie Sakson provided important guidance and help in editing and preparing this book for publication. We appreciate others who read drafts of parts or all of the book manuscript and provided helpful suggestions, including Abigail McBride, Katelyn Cranney, John Bachmann, Levi Severson, Sam Rushforth, Devin Pope, Mark Showalter, ViAnn Prestwich, Meredith Murdock, and others. Anonymous peer reviewers gave us thoughtful and constructive comments and suggestions.

We are grateful to our many wonderful and remarkable colleagues, including research collaborators, research assistants, mentors, and students. We appreciate various funding agencies, institutes, and our universities that have funded and supported our research over the last four decades. Finally, we recognize and are grateful for dedicated administrative assistants and staff who made it possible to conduct this work.

NOTES

PROLOGUE

1. C. Arden Pope, "Respiratory Disease Associated with Community Air Pollution and a Steel Mill, Utah Valley," *American Journal of Public Health* 79, no. 5 (1989): 623–628.

2. Brook Adams, "Illnesses Blamed on Virus, Not Pollution. Doctor Hired by Geneva to Analyze Data Disputes Findings of BYU Professor," *Deseret News*, September 2, 1989; see also Sheridan R. Hansen, "Geneva Says Respiratory Study Results Are Not Valid," *Deseret News*, December 12, 1988.

3. Douglas W. Dockery, Frank E. Speizer, Daniel O. Stram, James H. Ware, John D. Spengler, and Benjamin G. Ferris, "Effects of Inhalable Particles on Respiratory Health of Children," *American Review of Respiratory Disease* 139, no. 3 (1989): 587–594.

4. Janet Raloff, "Smallest Aerosol Pollutants Linked to Disease," *Science News*, May 135 (1989): 277.

CHAPTER 1

1. Claire Marshall, "Rosamund Adoo-Kissi-Debrah: Did Air Pollution Kill My Daughter?," *BBC News*, November 29, 2020, https://www.bbc.com/news/stories-55106501.

2. Sandra Laville, "Ella Kissi-Debrah: How a Mother's Fight for Justice May Help Prevent Other Air Pollution Deaths," *Guardian*, December 16, 2020, https://www.theguardian.com/environment/2020/dec/16/ella-kissi-debrah-mother-fight-justice-air-pollution-death.

3. Philip Barlow, "Regulation 28: Report to Prevent Future Deaths Following the Death of Ella Adoo Kissi-Debrah," Judiciary of England and Wales, April 20, 2021,

Annex A, https://www.judiciary.uk/wp-content/uploads/2021/04/Ella-Kissi-Debrah -2021-0113-1.pdf.

4. The Ella Roberta Foundation, "Clean Air for All." https://ellaroberta.org.

5. Joseph Bauman, "Effects of Pollution Range from Myriad Health Woes to Death," *Deseret News*, January 27, 1994, https://www.deseret.com/1994/1/27/19088914 /effects-of-pollution-range-from-myriad-health-woes-to-death.

6. C. Arden Pope, "Respiratory Disease Associated with Community Air Pollution and a Steel Mill, Utah Valley," *American Journal of Public Health* 79, no. 5 (1989): 623–628; C. Arden Pope, Douglas W. Dockery, John D. Spengler and Mark E. Raizenne, "Respiratory Health and PM$_{10}$ Pollution: A Daily Time Series Analysis," *American Review of Respiratory Disease* 144, no. 3.1 (1991): 668–674; C. Arden Pope and Douglas W. Dockery, "Acute Health Effects of PM$_{10}$ Pollution on Symptomatic and Asymptomatic Children," *American Review of Respiratory Disease* 145, no. 5 (1992): 1123–1128.

7. Aaron J. Cohen, Michael Brauer, Richard Burnett, H. Ross Anderson, Joseph Frostad, Kara Estep, Kalpana Balakrishnan, et al., "Estimates and 25-Year Trends of the Global Burden of Disease Attributable to Ambient Air Pollution: An Analysis of Data from the Global Burden of Diseases Study 2015," *Lancet* 389, no. 10082 (2017): 1907–1918; GBD 2021 Risk Factors Collaborators, "Global Burden and Strength of Evidence for 88 Risk Factors in 204 Countries and 811 Subnational Locations, 1990– 2021: A Systematic Analysis for the Global Burden of Disease Study 2021," *Lancet* 403, no. 10440 (2024): 2162–2024

8. David Michaels, *Doubt Is Their Product: How Industry's Assault on Science Threatens Your Health* (Oxford: Oxford University Press, 2008); see also David Michaels, *The Triumph of Doubt: Dark Money and the Science of Deception* (Oxford: Oxford University Press, 2020).

9. Naomi Oreskes and Erik M. Conway, *Merchants of Doubt: How a Handful of Scientists Obscured the Truth on Issues from Tobacco Smoke to Global Warming* (New York: Bloomsbury Press, 2010).

10. Ido Efrati, "Plaque Found on 400,000-Year-Old Teeth Shows Caves Had Dirty Air," *Haaretz*, June 19, 2015, https://www.haaretz.com/archaeology/2015-06-19/ty -article/.premium/findings-from-400-000-year-old-plaque/0000017f-df0b-d856-a37f -ffcb13fc0000; John D. Spengler and Ken Sexton, "Indoor Air Pollution: A Public Health Perspective," *Science* 221, no. 4605 (1983): 9–17.

11. Peter Brimblecombe, *The Big Smoke: A History of Air Pollution in London since Medieval Times* (London: Routledge, 1987).

12. Jon Krakauer, *Into Thin Air: A Personal Account of the Mt. Everest Disaster* (New York: Anchor Books, 1997), 54.

13. U.S. EPA, Particulate Matter (PM) Air Quality Standards, https://www.epa.gov /naaqs/particulate-matter-pm-air-quality-standards; WHO, "Ambient (Outdoor) Air Pollution: Fact Sheet, World Health Organization," September 22, 2021, https://

www.who.int/news-room/fact-sheets/detail/ambient-(outdoor)-air-quality-and
-health.

14. Robert D. Brook, Sanjay Rajagopalan, C. Arden Pope, Jeffrey R. Brook, Aruni
Bhatnagar, Ana V. Diez-Roux, Fernando Holguin, et al., "Particulate Matter Air Pollution and Cardiovascular Disease," *Circulation* 121, no. 21 (2010): 2331–2378.

CHAPTER 2

1. J. Firket, "Fog along the Meuse Valley," *Transactions of the Faraday Society* 32
(1936): 1192, https://doi.org/10.1039/tf9363201192.

2. Benoit Nemery, Peter H. M. Hoet and Abderrahim Nemmar, "The Meuse Valley
Fog of 1930: An Air Pollution Disaster," *Lancet* 357, no. 9257 (2001): 704–708, 707.

3. Firket, "Fog along the Muese Valley," 1192.

4. Lynne Page Snyder, "'The Death-Dealing Smog over Donora, Pennsylvania':
Industrial Air Pollution, Public Health Policy, and the Politics of Expertise, 1948–
1949," *Environmental History Review* 18, no. 1 (1994): 117–139.

5. Berton Roueché, *Eleven Blue Men, and Other Narratives of Medical Detection* (Boston:
Little, Brown, 1953), 201.

6. Roueché, *Eleven Blue Men*, 203.

7. Roueché, *Eleven Blue Men*, 204–205; see also Andy McPhee, "Saving Donorans on
a Deadly Night," February 21, 2018, https://andymcphee.com/2018/02/21/saving
-donorans-on-a-deadly-night.

8. Helmuth Herman Schrenk, H. Heimann, G. D. Clayton, W. M. Gafafer, and H.
Wexler, *Air Pollution in Donora, PA: Epidemiology of the Unusual Smog Episode of October 1948* (Washington, DC: U.S. Government Printing Office, 1949).

9. Antonio Ciocco and Donovan J. Thompson, "A Follow-Up of Donora Ten Years
After: Methodology and Findings," *American Journal of Public Health and the Nation's
Health* 51, no. 2 (1961): 155–164.

10. Snyder, "The Death-Dealing Smog," 132.

11. Kate Winkler Dawson, *Death in the Air* (New York: Hachette Books, 2017), 13.

12. Peter Brimblecombe, *The Big Smoke: A History of Air Pollution in London since
Medieval Times* (London: Routledge, 1987).

13. CwnEnvironment, "Great Smog of London Documentary, Part III," YouTube
video, February 15, 2012, 09:00, https://www.youtube.com/watch?v=cEK7POV8KSk.

14. W. P. D. Logan, "Mortality in London Fog Incident, 1952," *Lancet* 1, no. 6755
(1953): 336–338.

15. Michelle L. Bell and Devra L. Davis, "Reassessment of the Lethal London Fog of
1952: Novel Indicators of Acute and Chronic Consequences of Acute Exposure to Air
Pollution," *Environmental Health Perspectives* 109, no. 3 (2001): 389–394; Michelle L.
Bell, Devra L. Davis, and Tony Fletcher, "A Retrospective Assessment of Mortality

from the London Smog Episode of 1952: The Role of Influenza and Pollution," *Environmental Health Perspectives* 112, no. 1 (2004): 6–8.

16. U.S. EPA, Evolution of the Clean Air Act, https://www.epa.gov/clean-air-act-overview/evolution-clean-air-act; U.S. EPA, Particulate Matter (PM) Air Quality Standards, https://www.epa.gov/naaqs/particulate-matter-pm-air-quality-standards.

17. John Bachmann, "Will the Circle Be Unbroken: A History of the U.S. National Ambient Air Quality Standards," *Journal of the Air & Waste Management Association* 57, no. 6 (2007): 652–697.

CHAPTER 3

1. C. Arden Pope, "Respiratory Disease Associated with Community Air Pollution and a Steel Mill, Utah Valley," *American Journal of Public Health* 79, no. 5 (1989): 623–628.

2. John H. Cushman, "Utah Mill Lies at Heart of Fight for Air Pollution Limits," *New York Times*, January 21, 1997, https://www.nytimes.com/1997/01/21/us/utah-mill-lies-at-heart-of-fight-for-air-pollution-limits.html.

3. S. H. Lamm, T. A. Hall, A. Engel, F. H. Rueter and L. D. White, "PM_{10} Particulates: Are They the Major Determinant of Pediatric Respiratory Admissions in Utah County, Utah (1985–1989)?," *Annals of Occupational Hygiene* 38, Inhaled particles VII (1994): 969–972.

4. Public Relations Society of America, 1990, "Geneva Steel $2 Bill Campaign," A Silver-Anvil Award Entry, Inventory No. 6BW-9003A37.

5. Lyndsey Stram, "The Changing Utah County Economy," Department of Workforce Services, May 31, 2021, https://jobs.utah.gov/blog/post/2021/05/31/the-changing-utah-county-economy.

6. Amy Feldman, "Silicon Slopes vs. Silicon Valley: Four Tech Unicorns, Thousands of Startups, No Frenzy," *Forbes*, April 3, 2017, https://www.forbes.com/sites/amy-feldman/2017/04/03/silicon-slopes-vs-silicon-valley-four-tech-unicorns-thousands-of-startups-no-frenzy/?sh=6d07b2273922.

7. Davis Bell, "Utah's Highly Successful Tech Scene Highlights the Importance of Tech Ecosystems," *Forbes*, May 27, 2022, https://www.forbes.com/sites/forbestech-council/2022/05/27/utahs-highly-successful-tech-scene-highlights-the-importance-of-tech-ecosystems/?sh=5b0a2d6a5b51.

8. C. Arden Pope, "Respiratory Hospital Admissions Associated with PM_{10} Pollution in Utah, Salt Lake, and Cache Valleys," *Archives of Environmental Health: An International Journal* 46, no. 2 (1991): 90–97.

9. Michael R. Ransom and C. Arden Pope, "Elementary School Absences and PM_{10} Pollution in Utah Valley," *Environmental Research* 58, nos. 1–2 (1992): 204–219.

10. Mark W. Frampton, Andrew J. Ghio, James M. Samet, Johnny L. Carson, Jacqueline D. Carter, and Robert B. Devlin, "Effects of Aqueous Extracts of PM_{10} Filters from the Utah Valley on Human Airway Epithelial Cells," *American Journal*

of Physiology—Lung Cellular and Molecular Physiology 277, no. 5 (1999): 960–967; Andrew J. Ghio and Robert B. Devlin, "Inflammatory Lung Injury after Bronchial Instillation of Air Pollution Particles," *American Journal of Respiratory and Critical Care Medicine* 164, no. 4 (2001): 704–708; Janice A. Dye, James R. Lehmann, John K. McGee, Darrell W. Winsett, Allen D. Ledbetter, Jeffrey I. Everitt, Andrew J. Ghio, and Daniel L. Costa, "Acute Pulmonary Toxicity of Particulate Matter Filter Extracts in Rats: Coherence with Epidemiologic Studies in Utah Valley Residents," *Environmental Health Perspectives* 109 (2001): 395.

11. Douglas W. Dockery, C. S. Berkey, J. H. Ware, F. E. Speizer, and B. G. Ferris, "Distribution of Forced Vital Capacity and Forced Expiratory Volume in One Second in Children 6 to 11 Years of Age," *American Review of Respiratory Disease* 128, no. 3 (1983): 405–412; Douglas W. Dockery, James H. Ware, Benjamin G. Ferris, David S. Glicksberg, Martha E. Fay, Avron Spiro, and Frank E. Speizer, "Distribution of Forced Expiratory Volume in One Second and Forced Vital Capacity in Healthy, White, Adult Never-Smokers in Six U.S. Cities," *American Review of Respiratory Disease* 131, no. 4 (1985): 511–520.

12. Douglas W. Dockery, Frank E. Speizer, Daniel O. Stram, James H. Ware, John D. Spengler, and Benjamin G. Ferris, "Effects of Inhalable Particles on Respiratory Health of Children," *American Review of Respiratory Disease* 139, no. 3 (1989): 587–594.

13. C. Arden Pope, "Respiratory Disease Associated with Community Air Pollution and a Steel Mill, Utah Valley," *American Journal of Public Health* 79, no. 5 (1989): 623–628.

14. C. Arden Pope, Douglas W. Dockery, John D. Spengler, and Mark E. Raizenne, "Respiratory Health and PM_{10} Pollution: A Daily Time Series Analysis," *American Review of Respiratory Disease* 144, no. 3.1 (1991): 668–674; C. Arden Pope and Douglas W. Dockery, "Acute Health Effects of PM_{10} Pollution on Symptomatic and Asymptomatic Children," *American Review of Respiratory Disease* 145, no. 5 (1992): 1123–1128.

15. G. Hoek, D. W. Dockery, A. Pope, L. Neas, W. Roemer, and B. Brunekreef, "Association between PM_{10} and Decrements in Peak Expiratory Flow Rates in Children: Reanalysis of Data from Five Panel Studies," *European Respiratory Journal* 11, no. 6 (1998): 1307–1311; Diane R. Gold, Andrew I. Damokosh, C. Arden Pope, Douglas W. Dockery, William F. McDonnell, Paulina Serrano, Armando Retama, and Margarita Castillejos, "Particulate and Ozone Pollutant Effects on the Respiratory Function of Children in Southwest Mexico City," *Epidemiology* 10, no. 1 (1999): 8–16; C. Arden Pope and Richard E. Kanner, "Acute Effects of PM_{10} Pollution on Pulmonary Function of Smokers with Mild to Moderate Chronic Obstructive Pulmonary Disease," *American Review of Respiratory Disease* 147, no. 6.1 (1993): 1336–1340.

16. Joel Schwartz and Allan Marcus, "Mortality and Air Pollution in London: A Time Series Analysis," *American Journal of Epidemiology* 131, no. 1 (1990): 185–194.

17. Bart Ostro, "A Search for a Threshold in the Relationship of Air Pollution to Mortality: A Reanalysis of Data on London Winters," *Environmental Health Perspectives* 58 (1984): 397–399.

18. Joel Schwartz and Douglas W. Dockery, "Increased Mortality in Philadelphia Associated with Daily Air Pollution Concentrations," *American Review of Respiratory Disease* 145, no. 3 (1992): 600–604.

19. David Fairley, "The Relationship of Daily Mortality to Suspended Particulates in Santa Clara County, 1980–1986," *Environmental Health Perspectives* 89 (1990): 159–168.

20. Schwartz and Dockery, "Increased Mortality in Philadelphia"; Joel Schwartz and Douglas W. Dockery, "Particulate Air Pollution and Daily Mortality in Steubenville, Ohio," *American Journal of Epidemiology* 135, no. 1 (1992): 12–19; C. Arden Pope, Joel Schwartz, and Michael R. Ransom, "Daily Mortality and PM_{10} Pollution in Utah Valley," *Archives of Environmental Health: An International Journal* 47, no. 3 (1992): 211–217; Douglas W. Dockery, Joel Schwartz, and John D. Spengler, "Air Pollution and Daily Mortality: Associations with Particulates and Acid Aerosols," *Environmental Research* 59, no. 2 (1992): 362–373; Joel Schwartz, "Air Pollution and Daily Mortality in Birmingham, Alabama," *American Journal of Epidemiology* 137, no. 10 (1993): 1136–1147.

21. Robert. F. Phalen and R. O. McClellan, "PM-10 Research Needs," *Inhalation Toxicology* 7, no. 5 (1995): 773–779.

22. Robert F. Phalen and Jeffrey S. Lee, "Research Needs Relating to Health Effects of Exposure to Low Levels of Airborne Particulate Matter," *Applied Occupational and Environmental Hygiene* 13, no. 6 (1998): 352–355.

23. Suresh H. Moolgavkar and E. Georg Luebeck, "A Critical Review of the Evidence on Particulate Air Pollution and Mortality," *Epidemiology* 7, no. 4 (1996): 420–428; Frederick W. Lipfert and Ronald E. Wyzga, "Air Pollution and Mortality: The Implications of Uncertainties in Regression Modeling and Exposure Measurement," *Journal of the Air & Waste Management Association* 47, no. 4 (1997): 517–523.

24. C. Arden Pope and Laurence S. Kalkstein, "Synoptic Weather Modeling and Estimates of the Exposure-Response Relationship between Daily Mortality and Particulate Air Pollution," *Environmental Health Perspectives* 104, no. 4 (1996): 414–420.

25. Jonathan Samet, Scott Zeger, Julia Kelsall, Jing Xu, and Laurence Kalkstein, "Does Weather Confound or Modify the Association of Particulate Air Pollution with Mortality?," *Environmental Research* 77, no. 1 (1998): 9–19.

26. Health Effects Institute, "Particulate Air Pollution and Daily Mortality: Replication and Validation of Selected Studies," Phase I Report of the Particle Epidemiology Evaluation Project (Boston: Health Effects Institute, 1995).

27. Jonathon Samet, Francesca Dominici, Frank C. Curriero, Ivan Coursac, and Scott L. Zeger, "Fine Particulate Air Pollution and Mortality in 20 U.S. Cities, 1987–1994," *New England Journal of Medicine* 343, no. 24 (2000): 1742–1749; Francesca Dominici, Aidan McDermott, Scott L. Zeger, and Jonathan M. Samet, "National

Maps of the Effects of Particulate Matter on Mortality: Exploring Geographical Variation," *Environmental Health Perspectives* 111, no. 1 (2003): 39–44; Michelle L. Bell, Jonathan M. Samet, and Francesca Dominici, "Time-Series Studies of Particulate Matter," *Annual Review of Public Health* 25, no. 1 (2004): 247–280; Francesca Dominici, Roger D. Peng, Keita Ebisu, Scott L. Zeger, Jonathan M. Samet, and Michelle L. Bell, "Does the Effect of PM_{10} on Mortality Depend on PM Nickel and Vanadium Content? A Reanalysis of the NMMAPS Data," *Environmental Health Perspectives* 115, no. 12 (2007): 1701–1703; R. D. Peng, F. Dominici, R. Pastor-Barriuso, S. L. Zeger, and J. M. Samet, "Seasonal Analyses of Air Pollution and Mortality in 100 US Cities," *American Journal of Epidemiology* 161, no. 6 (2005): 585–594.

28. Klea Katsouyanni, Giota Touloumi, Evangelia Samoli, Alexandros Gryparis, Alain Le Tertre, Yannis Monopolis, Giuseppe Rossi, et al., "Confounding and Effect Modification in the Short-Term Effects of Ambient Particles on Total Mortality: Results from 29 European Cities within the APHEA2 Project," *Epidemiology* 12, no. 5 (2001): 521–531.

29. K. Katsouyanni, J. M. Samet, H. R. Anderson, R. Atkinson, et al., "Air Pollution and Health: A European and North American Approach (APHENA)," Research Report, *Health Effects Institute* 142 (2009): 5–90.

30. C. Liu, R. Chen, F. Sera, et al., "Ambient Air Pollution and Mortality in 652 Cities," *New England Journal of Medicine* 381, no. 21 (2019): 2072–2075.

31. John R. Balmes, "Do We Really Need Another Time-Series Study of the PM2.5–Mortality Association?," *New England Journal of Medicine* 381, no. 8 (2019): 774–776.

32. R. W. Atkinson, S. Kang, H. R. Anderson, I. C. Mills, and H. A. Walton, "Epidemiological Time Series Studies of $PM_{2.5}$ and Daily Mortality and Hospital Admissions: A Systematic Review and Meta-Analysis," *Thorax* 69, no. 7 (2014): 660–665.

33. Francesca Dominici, Roger D. Peng, Michelle L. Bell, Luu Pham, Aidan McDermott, Scott L. Zeger, and Jonathan M. Samet, "Fine Particulate Air Pollution and Hospital Admission for Cardiovascular and Respiratory Diseases," *JAMA* 295, no. 10 (2006): 1127–1134, 1127.

34. Malcolm Maclure, "The Case-Crossover Design: A Method for Studying Transient Effects on the Risk of Acute Events," *American Journal of Epidemiology* 133, no. 2 (1991): 144–153.

35. Murray A. Mittleman, Malcolm Maclure, Geoffrey H. Tofler, Jane B. Sherwood, Robert J. Goldberg, and James E. Muller, "Triggering of Acute Myocardial Infarction by Heavy Physical Exertion—Protection against Triggering by Regular Exertion," *New England Journal of Medicine* 329, no. 23 (1993): 1677–1683; Murray A. Mittleman, Malcolm Maclure, Jane B. Sherwood, Richard P. Mulry, Geoffrey H. Tofler, Sue C. Jacobs, Richard Friedman, Herbert Benson, and James E. Muller, "Triggering of Acute Myocardial Infarction Onset by Episodes of Anger," *Circulation* 92, no. 7 (1995): 1720–1725; James E. Muller, Murray A. Mittleman, Malcolm Maclure, et al., "Triggering Myocardial Infarction by Sexual Activity," *JAMA* 275, no. 18 (1996): 1405; Murray A. Mittleman, David Mintzer, Malcolm Maclure, Geoffrey H. Tofler, Jane B. Sherwood, and James E. Muller, "Triggering of Myocardial Infarction by Cocaine,"

Circulation 99, no. 21 (1999): 2737–2741; Murray A. Mittleman, Rebecca A. Lewis, Malcolm Maclure, Jane B. Sherwood, and James E. Muller, "Triggering Myocardial Infarction by Marijuana," *Circulation* 103, no. 23 (2001): 2805–2809.

36. L. M. Neas, J. Schwartz, and D. Dockery, "A Case-Crossover Analysis of Air Pollution and Mortality in Philadelphia," *Environmental Health Perspectives* 107, no. 8 (1999): 629–631; J. T. Lee and J. Schwartz, "Reanalysis of the Effects of Air Pollution on Daily Mortality in Seoul, Korea: A Case-Crossover Design," *Environmental Health Perspectives* 107, no. 8 (1999): 633–636; Joel Schwartz, "The Effects of Particulate Air Pollution on Daily Deaths: A Multi-City Case Crossover Analysis," *Occupational and Environmental Medicine* 61, no. 12 (2004): 956–961.

37. Annette Peters, Douglas W. Dockery, James E. Muller, and Murray A. Mittleman, "Increased Particulate Air Pollution and the Triggering of Myocardial Infarction," *Circulation* 103, no. 23 (2001): 2810–2815.

38. Holly Janes, Lianne Sheppard, and Thomas Lumley, "Case-Crossover Analyses of Air Pollution Exposure Data," *Epidemiology* 16, no. 6 (2005): 717–726; Holly Janes, Lianne Sheppard, and Thomas Lumley, "Overlap Bias in the Case-Crossover Design, with Application to Air Pollution Exposures," *Statistics in Medicine* 24, no. 2 (2005): 285–300.

39. Qian Di, Lingzhen Dai, Yun Wang, Antonella Zanobetti, Christine Choirat, Joel D. Schwartz, and Francesca Dominici, "Association of Short-Term Exposure to Air Pollution with Mortality in Older Adults," *JAMA* 318, no. 24 (2017): 2446–2456.

40. C. Arden Pope, "Mortality and Air Pollution: Associations Persist with Continued Advances in Research Methodology," *Environmental Health Perspectives* 107, no. 8 (1999): 613.

41. Benjamin D. Horne, Jeffrey L. Anderson, Jerry M. John, Aaron Weaver, Tami L. Bair, Kurt R. Jensen, Dale G. Renlund, and Joseph B. Muhlestein, "Which White Blood Cell Subtypes Predict Increased Cardiovascular Risk?," *Journal of the American College of Cardiology* 45, no. 10 (2005): 1638–1643.

42. C. Arden Pope, Joseph B. Muhlestein, Heidi T. May, Dale G. Renlund, Jeffrey L. Anderson, and Benjamin D. Horne, "Ischemic Heart Disease Events Triggered by Short-Term Exposure to Fine Particulate Air Pollution," *Circulation* 114, no. 23 (2006): 2443–2448.

43. C. Arden Pope, Dale G. Renlund, Abdallah G. Kfoury, Heidi T. May, and Benjamin D. Horne, "Relation of Heart Failure Hospitalization to Exposure to Fine Particulate Air Pollution," *American Journal of Cardiology* 102, no. 9 (2008): 1230–1234; C. Arden Pope, Joseph B. Muhlestein, Jeffrey L. Anderson, John B. Cannon, Nicholas M. Hales, Kent G. Meredith, Viet Le, and Benjamin D. Horne, "Short-Term Exposure to Fine Particulate Matter Air Pollution Is Preferentially Associated with the Risk of ST-Segment Elevation Acute Coronary Events," *Journal of the American Heart Association* 4, no. 12 (2015); 4:e002506; Benjamin D. Horne, Elizabeth A. Joy, Michelle G. Hofmann, Per H. Gesteland, John B. Cannon, Jacob S. Lefler, Denitza P. Blagev, et al., "Short-Term Elevation of Fine Particulate Matter Air Pollution and

Acute Lower Respiratory Infection," *American Journal of Respiratory and Critical Care Medicine* 198, no. 6 (2018): 759–766.

44. Blake Gardner, Frederick Ling, Philip K. Hopke, Mark W. Frampton, Mark J. Utell, Wojciech Zareba, Scott J. Cameron, et al., "Ambient Fine Particulate Air Pollution Triggers ST-Elevation Myocardial Infarction, but Not Non-ST Elevation Myocardial Infarction: A Case-Crossover Study," *Particle and Fibre Toxicology* 11, no. 1 (2014), https://doi.org/10.1186/1743-8977-11-1.

45. Hazrije Mustafić, Patricia Jabre, Christophe Caussin, Mohammad H. Murad, Sylvie Escolano, Muriel Tafflet, Marie-Cécile Périer, et al., "Main Air Pollutants and Myocardial Infarction," *JAMA* 307, no. 7 (2012): 713–721.

46. R. Chen, Y. Y. Jiang, J. Hu, et al., "Hourly Air Pollutants and Acute Coronary Syndrome Onset in 1.29 Million," *Circulation* 145 (2023): 1749–1760.

CHAPTER 4

1. Lester B. Lave and Eugene P. Seskin, "Air Pollution and Human Health," *Science* 169, no. 3947 (1970): 723–733.

2. L. B. Lave, E. P. Seskin, and M. J. Chappie, *Air Pollution and Human Health* (Baltimore: Johns Hopkins University Press, 1977).

3. Frederick W. Lipfert, "Air Pollution and Mortality: Specification Searches Using SMSA-Based Data," *Journal of Environmental Economics and Management* 3 (1984): 208–243; J. Evans, T. Tosteson, and P. Kinney, "Cross-Sectional Mortality Studies and Air Pollution Risk Assessment," *Environment International* 10, no. 1 (1984): 55–83; Frederick W. Lipfert and Ronald E. Wyzga, "Air Pollution and Mortality: The Implications of Uncertainties in Regression Modeling and Exposure Measurement," *Journal of the Air & Waste Management Association* 47, no. 4 (1997): 517–523; Robert Mendelsohn and Guy Orcutt, "An Empirical Analysis of Air Pollution Dose-Response Curves," *Journal of Environmental Economics and Management* 6, no. 2 (1979): 85–106.

4. Haluk Ozkaynak and George D. Thurston, "Associations between 1980 U.S. Mortality Rates and Alternative Measures of Airborne Particle Concentration," *Risk Analysis* 7, no. 4 (1987): 449–461.

5. Evans, Tosteson, and Kinney, "Cross-Sectional Mortality Studies and Air Pollution Risk."

6. B. E. Brown, "The Environmental Protection Agency's Research Program with Primary Emphasis on the Community Health and Environmental Surveillance System (CHESS): An Investigative Report" (Washington, DC: U.S. Government Printing Office, 1976).

7. Douglas W. Dockery, C. Arden Pope, Xiping Xu, John D. Spengler, James H. Ware, Martha E. Fay, Benjamin G. Ferris, and Frank E. Speizer, "An Association between Air Pollution and Mortality in Six U.S. Cities," *New England Journal of Medicine* 329, no. 24 (1993): 1753–1759.

8. C. Arden Pope, Michael J. Thun, Mohan M. Namboodiri, Douglas W. Dockery, John S. Evans, Frank E. Speizer, and Clark W. Heath, "Particulate Air Pollution as

a Predictor of Mortality in a Prospective Study of U.S. Adults," *American Journal of Respiratory and Critical Care Medicine* 151, no. 3.1 (1995): 669–674.

9. Douglas W. Dockery and C. Arden Pope, "Authors' Reply to Letter of S. H. Moolgavkar," *New England Journal of Medicine* 330 (1994): 1238.

10. American Lung Association v. Browner, 884 F. Supp. 345 (D. Ariz. 1994).

11. Whitman v. American Trucking Associations, Inc., 531 U.S. 457 (2001).

12. Jocelyn Kaiser, "Showdown over Clean Air Science," *Science* 277, no. 5325 (1997): 466–469.

13. D. Krewski, R. T. Burnett, M. S. Goldberg, K. Hoover, J. Siemiatycki, M. Jarret, M. Abrahamowicz, and W. H. White, *Reanalysis of the Harvard Six Cities Study and the American Cancer Society Study of Particulate Air Pollution and Mortality: A Special Report of the Institute's Particle Epidemiology Reanalysis Project* (Cambridge, MA: Health Effects Institute, 2000).

14. Krewski et al., *Reanalysis of the Harvard Six Cities Study.*

15. Health Effects Institute, "New Analysis Confirms Results of Key Studies of Particles and Mortality," press release, July 26, 2000.

16. Krewski et al., *Reanalysis of the Harvard Six Cities Study.*

17. Krewski et al., *Reanalysis of the Harvard Six Cities Study,* 275.

18. C. Arden Pope III, R. T. Burnett, M. J. Thun, E. E. Calle, D. Krewski, K. Ito, and G. D. Thurston, "Lung Cancer, Cardiopulmonary Mortality, and Long-Term Exposure to Fine Particulate Air Pollution," *JAMA* 287, no. 9 (2002): 1132–1141.

19. C. Arden Pope, Richard T. Burnett, George D. Thurston, Michael J. Thun, Eugenia E. Calle, Daniel Krewski, and John J. Godleski, "Cardiovascular Mortality and Long-Term Exposure to Particulate Air Pollution," *Circulation* 109, no. 1 (2004): 71–77.

20. C. Arden Pope, Michelle C. Turner, Richard T. Burnett, Michael Jerrett, Susan M. Gapstur, W. Ryan Diver, Daniel Krewski, and Robert D. Brook, "Relationships between Fine Particulate Air Pollution, Cardiometabolic Disorders, and Cardiovascular Mortality," *Circulation Research* 116, no. 1 (2015): 108–115.

21. Michelle Turner, Daniel Krewski, C. Arden Pope, Yue Chen, Susan M. Gapstur, and Michael J. Thun, "Long-Term Ambient Fine Particulate Matter Air Pollution and Lung Cancer in a Large Cohort of Never-Smokers," *American Journal of Respiratory and Critical Care Medicine* 184, no. 12 (2011): 1374–1381.

22. Michelle Turner, Aaron Cohen, Michael Jerrett, Susan M. Gapstur, W. Ryan Diver, C. Arden Pope, Daniel Krewski, Bernardo S. Beckerman, and Jonathan M. Samet, "Interactions between Cigarette Smoking and Fine Particulate Matter in the Risk of Lung Cancer Mortality in Cancer Prevention Study II," *American Journal of Epidemiology* 180, no. 12 (2016): 1145–1149.

23. Michael Jerrett, Richard T. Burnett, C. Arden Pope, Kazuhiko Ito, George Thurston, Daniel Krewski, Yuanli Shi, Eugenia Calle, and Michael Thun, "Long-Term

Ozone Exposure and Mortality," *New England Journal of Medicine* 360, no. 11 (2009): 1085–1095.

24. Kirk R. Smith, Michael Jerrett, H. Ross Anderson, Richard T. Burnett, Vicki Stone, Richard Derwent, Richard W. Atkinson, et al., "Public Health Benefits of Strategies to Reduce Greenhouse-Gas Emissions: Health Implications of Short-Lived Greenhouse Pollutants," *Lancet* 374, no. 9707 (2009): 2091–2103.

25. Michael Jerrett, Richard T. Burnett, Bernardo S. Beckerman, Michelle C. Turner, Daniel Krewski, George Thurston, Randall V. Martin, et al., "Spatial Analysis of Air Pollution and Mortality in California," *American Journal of Respiratory and Critical Care Medicine* 188, no. 5 (2013): 593–599.

26. Michael Jerrett, Michelle C. Turner, Bernardo S. Beckerman, C. Arden Pope, Aaron van Donkelaar, Randall V. Martin, Marc Serre, et al., "Comparing the Health Effects of Ambient Particulate Matter Estimated Using Ground-Based versus Remote Sensing Exposure Estimates," *Environmental Health Perspectives* 125, no. 4 (2017): 552–559.

27. George D. Thurston, Richard T. Burnett, Michelle C. Turner, Yuanli Shi, Daniel Krewski, Ramona Lall, Kazuhiko Ito, et al., "Ischemic Heart Disease Mortality and Long-Term Exposure to Source-Related Components of U.S. Fine Particle Air Pollution," *Environmental Health Perspectives* 124, no. 6 (2016): 785–794.

28. Francine Laden, Joel Schwartz, Frank E. Speizer, and Douglas W. Dockery, "Reduction in Fine Particulate Air Pollution and Mortality," *American Journal of Respiratory and Critical Care Medicine* 173, no. 6 (2006): 667–672.

29. Joel Schwartz, Brent Coull, Francine Laden, and Louise Ryan, "The Effect of Dose and Timing of Dose on the Association between Airborne Particles and Survival," *Environmental Health Perspectives* 115, no. 1 (2008): 64–69.

30. Johanna Lepeule, Francine Laden, Douglas Dockery, and Joel Schwartz, "Chronic Exposure to Fine Particles and Mortality: An Extended Follow-Up of the Harvard Six Cities Study from 1974 to 2009," *Environmental Health Perspectives* 120, no. 7 (2012): 965–970.

CHAPTER 5

1. Elaine Appleton Grant, "Prevailing Winds: A Decades-Long Fight to Bring Clean Air Standards in Line with Environmental Health Science Offers Lessons for Today," *Harvard Public Health*, fall 2012, https://www.hsph.harvard.edu/news/magazine/f12-six-cities-environmental-health-air-pollution.

2. J. E. Enstrom and Geoffrey C. Kabat, "Environmental Tobacco Smoke and Tobacco Related Mortality in a Prospective Study of Californians, 1960–1998," *BMJ* 326, no. 7398 (2003): 1057.

3. United States v. Philip Morris USA Inc., 2006, 449F. Supp. 2d 1. (D.D.C. 2006).

4. Elisa K. Tong and Stanton A. Glantz, "Tobacco Industry Efforts Undermining Linking Secondhand Smoke with Cardiovascular Disease," *Circulation* 116, no. 16 (2007): 1845–1854, 1848.

5. Jame E. Enstrom, "Fine Particulate Air Pollution and Total Mortality among Elderly Californians, 1973–2002," *Inhalation Toxicology* 17, no. 14 (2005): 803–816.

6. Bert Brunekreef and Gerard Hoek, "A Critique of 'Fine Particulate Air Pollution and Total Mortality among Elderly Californians, 1973–2002' by James E. Enstrom," *Inhalation Toxicology* 18, no. 7 (2006): 507–508.

7. James E. Enstrom, "Fine Particulate Matter and Total Mortality in Cancer Prevention Study Cohort Reanalysis," *Dose-Response* 15, no. 1 (2017): 155932581769334.

8. C. Arden Pope, Daniel Krewski, Susan M. Gapstur, Michelle C. Turner, Michael Jerrett, and Richard T. Burnett, "Fine Particulate Air Pollution and Mortality: Response to Enstrom's Reanalysis of the American Cancer Society Cancer Prevention Study II Cohort," *Dose-Response* 15, no. 4 (2017): 155932581774630.

9. Robert D. Langer, Emily White, Cora E. Lewis, Jane M. Kotchen, Susan L. Hendrix, and Maurizio Trevisan, "The Women's Health Initiative Observational Study: Baseline Characteristics of Participants and Reliability of Baseline Measures," *Annals of Epidemiology* 13, no. 9 (2003): S107–S121.

10. Writing Group for the Women's Health Initiative Investigators, "Risks and Benefits of Estrogen plus Progestin in Healthy Postmenopausal Women: Principal Results from the Women's Health Initiative Randomized Controlled Trial," *JAMA: The Journal of the American Medical Association* 288, no. 3 (2022): 321–333.

11. Roger A. Lobo, "Where Are We 10 Years after the Women's Health Initiative?," *Journal of Clinical Endocrinology & Metabolism*, 98, no. 5 (2013): 1771–1780.

12. Kristin A. Miller, David S. Siscovick, Lianne Sheppard, Kristen Shepherd, Jeffrey H. Sullivan, Garnet L. Anderson, and Joel D. Kaufman, "Long-Term Exposure to Air Pollution and Incidence of Cardiovascular Events in Women," *New England Journal of Medicine* 356, no. 5 (2007): 447–458.

13. William F. McDonnell, Naoi Nishino-Ishikawa, Floyd F. Petersen, Lie Hong Chen, and David E. Abbey, "Relationships of Mortality with the Fine and Coarse Fractions of Long-Term Ambient PM_{10} Concentrations in Nonsmokers," *Journal of Exposure Science & Environmental Epidemiology* 10, no. 5 (2000): 427–436; Robin C. Puett, Jaime E. Hart, Jeff D. Yanosky, Christopher Paciorek, Joel Schwartz, Helen Suh, Frank E. Speizer, and Francine Laden, "Chronic Fine and Coarse Particulate Exposure, Mortality, and Coronary Heart Disease in the Nurses' Health Study," *Environmental Health Perspectives* 117, no. 11 (2009): 1697–1701; Jaime E. Hart, Xiaomei Liao, Biling Hong, Robin C. Puett, Jeff D. Yanosky, Helen Suh, Marianthi-Anna Kioumourtzoglou, Donna Spiegelman, and Francine Laden, "The Association of Long-Term Exposure to $PM_{2.5}$ on All-Cause Mortality in the Nurses' Health Study and the Impact of Measurement-Error Correction," *Environmental Health* 14, no. 1 (2015), https://doi.org/10.1186/s12940-015-0027-6; Robin C. Puett, Jaime E. Hart, Helen Suh, Murray Mittleman, and Francine Laden, "Particulate Matter Exposures, Mortality, and Cardiovascular Disease in the Health Professionals Follow-Up Study," *Environmental Health Perspectives* 119, no. 8 (2011): 1130–1135; F. W. Lipfert, R. E. Wyzga, J. D. Baty, and J. P. Miller, "Traffic Density as a Surrogate Measure of Environmental Exposures in Studies of Air Pollution Health Effects: Long-Term Mortality

in a Cohort of US Veterans," *Atmospheric Environment* 40, no. 1 (2006): 154–169; Bart Ostro, Jianlin Hu, Debbie Goldberg, Peggy Reynolds, Andrew Hertz, Leslie Bernstein, and Michael J. Kleeman, "Associations of Mortality with Long-Term Exposures to Fine and Ultrafine Particles, Species and Sources: Results from the California Teachers Study Cohort," *Environmental Health Perspectives* 123, no. 6 (2015): 549–556; Scott Weichenthal, Paul J. Villeneuve, Richard T. Burnett, Aaron van Donkelaar, Randall V. Martin, Rena R. Jones, Curt T. DellaValle, Dale P. Sandler, Mary H. Ward, and Jane A. Hoppin, "Long-Term Exposure to Fine Particulate Matter: Association with Nonaccidental and Cardiovascular Mortality in the Agricultural Health Study Cohort," *Environmental Health Perspectives* 122, no. 6 (2014): 609–615; George D. Thurston, Jiyoung Ahn, Kevin R. Cromar, Yongzhao Shao, Harmony R. Reynolds, Michael Jerrett, Chris C. Lim, Ryan Shanley, Yikyung Park, and Richard B. Hayes, "Ambient Particulate Matter Air Pollution Exposure and Mortality in the NIH-AARP Diet and Health Cohort," *Environmental Health Perspectives* 124, no. 4 (2016): 484–490.

14. Nathan C. Coleman, Majid Ezzati, Julian D. Marshall, Allen L. Robinson, Richard T. Burnett, and C. Arden Pope, "Fine Particulate Matter Air Pollution and Mortality Risk among US Cancer Patients and Survivors," *JNCI Cancer Spectrum* 5, no. 1 (2021), https://doi.org/10.1093/jncics/pkab001.

15. Stacey F. Alexeeff, Kamala Deosaransingh, Noelle S. Liao, Stephen K. Van Den Eeden, Joel Schwartz, and Stephen Sidney, "Particulate Matter and Cardiovascular Risk in Adults with Chronic Obstructive Pulmonary Disease," *American Journal of Respiratory and Critical Care Medicine* 204, no. 2 (2021): 159–167.

16. Giulia Cesaroni, Chiara Badaloni, Claudio Gariazzo, Massimo Stafoggia, Roberto Sozzi, Marina Davoli, and Francesco Forastiere, "Long-Term Exposure to Urban Air Pollution and Mortality in a Cohort of More than a Million Adults in Rome," *Environmental Health Perspectives* 121, no. 3 (2013): 324–331; Rob Beelen, Gerard Hoek, Piet A. van den Brandt, R. Alexandra Goldbohm, Paul Fischer, Leo J. Schouten, Michael Jerrett, Edward Hughes, Ben Armstrong, and Bert Brunekreef, "Long-Term Effects of Traffic-Related Air Pollution on Mortality in a Dutch Cohort (NLCS-Air Study)," *Environmental Health Perspectives* 116, no. 2 (2008): 196–202; Paul H. Fischer, Marten Marra, Caroline B. Ameling, Gerard Hoek, Rob Beelen, Kees de Hoogh, Oscar Breugelmans, Hanneke Kruize, Nicole A. H. Janssen, and Danny Houthuijs, "Air Pollution and Mortality in Seven Million Adults: The Dutch Environmental Longitudinal Study (Duels)," *Environmental Health Perspectives* 123, no. 7 (2015): 697–704; Iain M. Carey, Richard W. Atkinson, Andrew J. Kent, Tjeerd van Staa, Derek G. Cook, and H. Ross Anderson, "Mortality Associations with Long-Term Exposure to Outdoor Air Pollution in a National English Cohort," *American Journal of Respiratory and Critical Care Medicine* 187, no. 11 (2013): 1226–1233; Malek Bentayeb, Verene Wagner, Morgane Stempfelet, Marie Zins, Marcel Goldberg, Mathilde Pascal, Sophie Larrieu, et al., "Association between Long-Term Exposure to Air Pollution and Mortality in France: A 25-Year Follow-Up Study," *Environment International* 85 (2015): 5–14; Ulla Arthur Hvidtfeldt, Mette Sørensen, Camilla Geels, Matthias Ketzel, Jibran Khan, Anne Tjønneland, Kim Overvad, Jørgen Brandt, and Ole Raaschou-Nielsen,

"Long-Term Residential Exposure to PM$_{2.5}$, PM$_{10}$, Black Carbon, NO$_2$, and Ozone and Mortality in a Danish Cohort," *Environment International* 123 (2019): 265–272.

17. Rob Beelen, Ole Raaschou-Nielsen, Massimo Stafoggia, Zorana Jovanovic Andersen, Gudrun Weinmayr, Barbara Hoffmann, Kathrin Wolf, et al., "Effects of Long-Term Exposure to Air Pollution on Natural-Cause Mortality: An Analysis of 22 European Cohorts within the Multicentre Escape Project," *Lancet* 383, no. 9919 (2014): 785–795; Rob Beelen, Massimo Stafoggia, Ole Raaschou-Nielsen, Zorana Jovanovic Andersen, Wei W. Xun, Klea Katsouyanni, Konstantina Dimakopoulou, et al., "Long-Term Exposure to Air Pollution and Cardiovascular Mortality," *Epidemiology* 25, no. 3 (2014): 368–378.

18. Bert Brunekreef, M. Strak, J. Chen, Z. J. Andersen, R. Atkinson, M. Bauwelinck, et al., "Mortality and Morbidity Effects of Long-Term Exposure to Low-Level PM$_{2.5}$, BC, NO$_2$, and O$_3$: An Analysis of European Cohorts in the ELAPSE Project," Research Report 208 (Boston: Health Effects Institute, 2021); Maciej Strak, Gudrun Weinmayr, Sophia Rodopoulou, Jie Chen, Kees de Hoogh, Zorana J. Andersen, Richard Atkinson, et al., "Long Term Exposure to Low Level Air Pollution and Mortality in Eight European Cohorts within the ELAPSE PROJECT: Pooled Analysis," *BMJ* 374, no. 1904 (2021), https://doi.org/10.1136/bmj.n1904.

19. Dan L. Crouse, Paul A. Peters, Perry Hystad, Jeffrey R. Brook, Aaron van Donkelaar, Randall V. Martin, Paul J. Villeneuve, et al., "Ambient PM$_{2.5}$, O$_3$, and NO$_2$ Exposures and Associations with Mortality over 16 Years of Follow-Up in the Canadian Census Health and Environment Cohort (CANCHEC)," *Environmental Health Perspectives* 123, no. 11 (2015): 1180–1186; Scott Weichenthal, Daniel L. Crouse, Lauren Pinault, Krystal Godri-Pollitt, Eric Lavigne, Greg Evans, Aaron van Donkelaar, Randall V. Martin, and Rick T. Burnett, "Oxidative Burden of Fine Particulate Air Pollution and Risk of Cause-Specific Mortality in the Canadian Census Health and Environment Cohort (CANCHEC)," *Environmental Research* 146 (2016): 92–99; Lauren Pinault, Michael Tjepkema, Daniel L. Crouse, Scott Weichenthal, Aaron van Donkelaar, Randall V. Martin, Michael Brauer, Hong Chen, and Richard T. Burnett, "Risk Estimates of Mortality Attributed to Low Concentrations of Ambient Fine Particulate Matter in the Canadian Community Health Survey Cohort," *Environmental Health* 15, no. 1 (2016), https://doi.org/10.1186/s12940-016-0111-6; Sabit Cakmak, Chris Hebbern, Lauren Pinault, Eric Lavigne, Jennifer Vanos, Dan Lawson Crouse, and Michael Tjepkema, "Associations between Long-Term PM$_{2.5}$ and Ozone Exposure and Mortality in the Canadian Census Health and Environment Cohort (CANCHEC), by Spatial Synoptic Classification Zone," *Environment International* 111 (2018): 200–211.

20. M. Brauer, J. R. Brook, T. Christidis, Y. Chu, D. L. Crouse, A. Erickson, P. Hystad, et. al, "Mortality-Air Pollution Associations in Low Exposure Environments (MAPLE)," Phase 2, Research Report 212 (Boston: Health Effects Institute, 2022).

21. Chit Ming Wong, Hak Kan Lai, Hilda Tsang, Thuan Quoc Thach, G. Neil Thomas, Kin Bong Lam, King Pan Chan, et al., "Satellite-Based Estimates of Long-Term Exposure to Fine Particles and Association with Mortality in Elderly Hong Kong Residents," *Environmental Health Perspectives* 123, no. 11 (2015): 1167–1172.

22. Peng Yin, Michael Brauer, Aaron Cohen, Richard T. Burnett, Jiangmei Liu, Yunning Liu, Ruiming Liang, et al., "Long-Term Fine Particulate Matter Exposure and Nonaccidental and Cause-Specific Mortality in a Large National Cohort of Chinese Men," *Environmental Health Perspectives* 125, no. 11 (2017): 117002.

23. Tiantian Li, Yi Zhang, Jiaonan Wang, Dandan Xu, Zhaoxue Yin, Huashuai Chen, Yuebin Lv, et al., "All-Cause Mortality Risk Associated with Long-Term Exposure to Ambient $PM_{2.5}$ in China: A Cohort Study," *Lancet Public Health* 3, no. 10 (2018): e470–e477.

24. Xueli Yang, Fengchao Liang, Jianxin Li, Jichun Chen, Fangchao Liu, Keyong Huang, Jie Cao, et al., "Associations of Long-Term Exposure to Ambient $PM_{2.5}$ with Mortality in Chinese Adults: A Pooled Analysis of Cohorts in the China-PAR Project," *Environment International* 138 (2020): 105589.

25. Mahshid Dehghan, Andrew Mente, Xiaohe Zhang, Sumathi Swaminathan, Wei Li, Viswanathan Mohan, Romaina Iqbal, et al., "Associations of Fats and Carbohydrate Intake with Cardiovascular Disease and Mortality in 18 Countries from Five Continents (PURE): A Prospective Cohort Study," *Lancet* 390, no. 10107 (2017): 2050–2062.

26. Perry Hystad, Andrew Larkin, Sumathy Rangarajan, Khalid F. AlHabib, Álvaro Avezum, Kevser Burcu Calik, Jephat Chifamba, et al., "Associations of Outdoor Fine Particulate Air Pollution and Cardiovascular Disease in 157 436 Individuals from 21 High-Income, Middle-Income, and Low-Income Countries (PURE): A Prospective Cohort Study," *Lancet Planetary Health* 4, no. 6 (2020): e235–e245.

27. C. Arden Pope and Douglas W. Dockery, "Health Effects of Fine Particulate Air Pollution: Lines That Connect," *Journal of the Air & Waste Management Association* 56, no. 6 (2006): 709–742.

28. Sverre Vedal, "Ambient Particles and Health: Lines That Divide," *Journal of the Air & Waste Management Association* 47, no. 5 (1997): 551–581.

29. Pope and Dockery, "Lines That Connect."

30. U.S. Senate Committee of Environment and Public Works, "Vitter, EPW Republicans Hold McCarthy to Promises of Increased Agency Transparency," press release, April 16, 2013, https://www.epw.senate.gov/public/index.cfm/2013/4/post-13c6cf9c-ff44-21d6-d579-3d81fca027b4.

31. Mark Drajem, "Two-Decade-Old Harvard Data Confounds U.S. EPA Nomination," *Bloomberg*, May 16, 2013, https://www.bloomberg.com/news/articles/2013-05-16/two-decade-old-harvard-data-confounds-u-s-epa-nomination.

32. U.S. Senate Committee of Environment and Public Works, "Vitter, EPW Republicans Hold McCarthy to Promises of Increased Agency Transparency" (see request 3).

33. Warren Cornwall, "Critics See Hidden Goal in EPA Data Access Rule," *Science* 360, no. 6388 (2018): 472–473, 472, https://doi.org/10.1126/science.360.6388.472.

34. Carolyn Kormann, "Scott Pruitt's Crusade against 'Secret Science' Could Be Disastrous for Public Health," *New Yorker*, April 26, 2018, https://www.newyorker

.com/science/elements/scott-pruitts-crusade-against-secret-science-could-be
-disastrous-for-public-health.

35. Sudip Parihk, "AAAS Statement: EPA 'Transparency Rule' Weakens the Use of Science in Policymaking," American Association for the Advancement of Science, March 18, 2020, https://www.aaas.org/sites/default/files/2020-03/EPA%20 %27Transparency%20Rule%27%20Weakens%20the%20Use%20of%20Science %20in%20Policymaking.pdf?adobe_mc=MCMID%3D81733012598554376 2707 87499290625651004%7CMCORGID%3D242B6472541199F70A4C98A6%2540 AdobeOrg%7CTS%3D1646773925.

36. H. Holden Thorp, Magdalena Skipper, Veronique Kiermer, May Berenbaum, Deborah Sweet, and Richard Horton, "Joint Statement on EPA Proposed Rule and Public Availability of Data," *Science* 366, no. 6470 (2019), https://www.science.org/ doi/10.1126/science.aba3197.

37. Thorp et al., "Joint Statement on EPA Proposed Rule," para. 2 and 3.

38. Envt'l Defense Fund v. EPA, D. Mont., No. 4:21-cv-00003, February 1, 2021.

39. Susan Cosier, "Clever Use of Public Data Could Sidestep New Rule," *Science* 360, no. 6388 (2018): 473.

40. Qian Di, Yan Wang, Antonella Zanobetti, Yun Wang, Petros Koutrakis, Christine Choirat, Francesca Dominici, and Joel D. Schwartz, "Air Pollution and Mortality in the Medicare Population," *New England Journal of Medicine* 376, no. 26 (2017): 2513–2522.

41. Jennifer D. Parker, Nataliya Kravets, and Ambarish Vaidyanathan, "Particulate Matter Air Pollution Exposure and Heart Disease Mortality Risks by Race and Ethnicity in the United States," *Circulation* 137, no. 16 (2018): 1688–1697.

42. C. Arden Pope, Jacob S. Lefler, Majid Ezzati, Joshua D. Higbee, Julian D. Marshall, Sun-Young Kim, Matthew Bechle, et al., "Mortality Risk and Fine Particulate Air Pollution in a Large, Representative Cohort of U.S. Adults," *Environmental Health Perspectives* 127, no. 7 (2019): 077007; Jacob S. Lefler, Joshua D. Higbee, Richard T. Burnett, Majid Ezzati, Nathan C. Coleman, Dalton D. Mann, Julian D. Marshall, et al., "Air Pollution and Mortality in a Large, Representative U.S. Cohort: Multiple-Pollutant Analyses, and Spatial and Temporal Decompositions," *Environmental Health* 18, no. 1 (2019), https://doi.org/10.1186/s12940-019-0544-9.

43. Nathan C. Coleman, Richard T. Burnett, Majid Ezzati, Julian D. Marshall, Allen L. Robinson, and C. Arden Pope, "Fine Particulate Matter Exposure and Cancer Incidence: Analysis of Seer Cancer Registry Data from 1992–2016," *Environmental Health Perspectives* 128, no. 10 (2020), https://doi.org/10.1289/ehp7246.

44. Coleman, Ezzati, Marshall, et al., "Fine Particulate Matter Air Pollution and Mortality Risk among US Cancer Patients and Survivors"; Carver J. Coleman, Ray A. Yeager, Zachari A. Pond, Daniel W. Riggs, Aruni Bhatnagar, and C. Arden Pope, "Mortality Risk Associated with Greenness, Air Pollution, and Physical Activity in a Representative U.S. Cohort," *Science of the Total Environment* 824 (2022): 153848.

45. This figure was produced by the authors using results from fifteen publications discussed and referenced in chapters 4 and 5: Douglas W. Dockery, C. Arden Pope,

Xiping Xu, John D. Spengler, James H. Ware, Martha E. Fay, Benjamin G. Ferris, and Frank E. Speizer, "An Association between Air Pollution and Mortality in Six U.S. Cities," *New England Journal of Medicine* 329, no. 24 (1993): 1753–1759; Johanna Lepeule, Francine Laden, Douglas Dockery, and Joel Schwartz, "Chronic Exposure to Fine Particles and Mortality: An Extended Follow-Up of the Harvard Six Cities Study from 1974 to 2009," *Environmental Health Perspectives* 120, no. 7 (2012): 965–970; C. Arden Pope, Michael J. Thun, Mohan M. Namboodiri, Douglas W. Dockery, John S. Evans, Frank E. Speizer, and Clark W. Heath, "Particulate Air Pollution as a Predictor of Mortality in a Prospective Study of U.S. Adults," *American Journal of Respiratory and Critical Care Medicine* 151, no. 3.1 (1995): 669–674; C. Arden Pope, Michelle C. Turner, Richard T. Burnett, Michael Jerrett, Susan M. Gapstur, W. Ryan Diver, Daniel Krewski, and Robert D. Brook, "Relationships between Fine Particulate Air Pollution, Cardiometabolic Disorders, and Cardiovascular Mortality," *Circulation Research* 116, no. 1 (2015): 108–115; Di, Wang, Zanobetti, Wang, et al., "Air Pollution and Mortality in the Medicare Population"; Pope, Lefler, Ezzati, Higbee, et al., "Mortality Risk and Fine Particulate Air Pollution in a Large, Representative Cohort of U.S. Adults"; Brauer, Brook, Christidis, Chu, et al., "Mortality-Air Pollution Associations in Low Exposure Environments (MAPLE)"; Brunekreef, Strak, Chen, Andersen, et al., "Mortality and Morbidity Effects of Long-Term Exposure to Low-Level $PM_{2.5}$, BC, NO_2, and O_3"; Wong, Lai, Tsang, Thach, et al., "Satellite-Based Estimates of Long-Term Exposure to Fine Particles and Association with Mortality in Elderly Hong Kong Residents"; Yin, Brauer, Cohen, Burnett, et al., "Long-Term Fine Particulate Matter Exposure and Nonaccidental and Cause-Specific Mortality in a Large National Cohort of Chinese Men"; Li, Zhang, Wang, Xu, et al., "All-Cause Mortality Risk Associated with Long-Term Exposure to Ambient $PM_{2.5}$ in China"; Yang, Liang, Li, Chen, et al., "Associations of Long-Term Exposure to Ambient $PM_{2.5}$ with Mortality in Chinese Adults"; Hystad, Larkin, Rangarajan, AlHabib, et al., "Associations of Outdoor Fine Particulate Air Pollution and Cardiovascular Disease in 157 436 Individuals from 21 High-Income, Middle-Income, and Low-Income Countries (PURE)"; Richard T. Burnett, Joseph V. Spadaro, George R. Garcia, and C. Arden Pope, "Designing Health Impact Functions to Assess Marginal Changes in Outdoor Fine Particulate Matter," *Environmental Research* 204 (2022): 112245; Jie Chen and Gerard Hoek, "Long-Term Exposure to PM and All-Cause and Cause-Specific Mortality: A Systematic Review and Meta-Analysis," *Environment International* 143 (2020): 105974.

46. Chen and Hoek, "Long-Term Exposure to PM and All-Cause and Cause-Specific Mortality"; Burnett, Spadaro, Garcia, and Pope, "Designing Health Impact Functions to Assess Marginal Changes in Outdoor Fine Particulate Matter."

CHAPTER 6

1. Jane Kay and Cheryl Katz, "Pollution, Poverty, and People of Color: Living with Industry," *Scientific American*, Special Report, June 4, 2012, para. 1, https://www.scientificamerican.com/article/pollution-poverty-people-color-living-industry.

2. Kay and Katz, "Pollution, Poverty, and People of Color," para. 5.

3. Kay and Katz, "Pollution, Poverty, and People of Color," para. 12.

4. Kyle J. Colonna, Petros Koutrakis, Patrick L. Kinney, Roger M. Cooke, and John S. Evans, "Mortality Attributable to Long-Term Exposure to Ambient Fine Particulate Matter: Insights from the Epidemiologic Evidence for Understudied Locations," *Environmental Science & Technology* 56, no. 11 (2022): 6799–6812.

5. Qian Di, Yan Wang, Antonella Zanobetti, Yun Wang, Petros Koutrakis, Christine Choirat, Francesca Dominici, and Joel D. Schwartz, "Air Pollution and Mortality in the Medicare Population," *New England Journal of Medicine* 376, no. 26 (2017): 2513–2522.

6. C. Arden Pope, Jacob S. Lefler, Majid Ezzati, Joshua D. Higbee, Julian D. Marshall, Sun-Young Kim, Matthew Bechle, et al., "Mortality Risk and Fine Particulate Air Pollution in a Large, Representative Cohort of U.S. Adults," *Environmental Health Perspectives* 127, no. 7 (2019): 077007.

7. Christopher W. Tessum, David A. Paolella, Sarah E. Chambliss, Joshua S. Apte, Jason D. Hill, and Julian D. Marshall, "PM 2.5 Polluters Disproportionately and Systemically Affect People of Color in the United States," *Science Advances* 7, no. 18 (2021), https://doi.org/10.1126/sciadv.abf4491.

8. University of Washington, "People of Color Hardest Hit by Air Pollution from Nearly all Sources," *University of Washington News*, April 28, 2021, para. 14, https://www.washington.edu/news/2021/04/28/people-of-color-hardest-hit-by-air-pollution-from-nearly-all-sources.

9. Abdulrahman Jbaily, Xiaodan Zhou, Jie Liu, Ting-Hwan Lee, Leila Kamareddine, Stéphane Verguet, and Francesca Dominici, "Air Pollution Exposure Disparities across U.S. Population and Income Groups," *Nature* 601, no. 7892 (2022): 228–233.

10. Haley M. Lane, Rachel Morello-Frosch, Julian D. Marshall, and Joshua S. Apte, "Historical Redlining Is Associated with Present-Day Air Pollution Disparities in U.S. Cities," *Environmental Science & Technology Letters* 9, no. 4 (2022): 345–350.

11. A. Hajat, C. Hsia, and M. S. O'Neill, "Socioeconomic Disparities and Air Pollution Exposure: A Global Review," *Current Environmental Health Reports* 2 (2015): 440–450.

12. Michelle Bell and Keita Ebisu, "Environmental Inequality in Exposures to Airborne Particulate Matter Components in the United States," *Environmental Health Perspectives* 120 (2012): 1699–1704.

13. David Reichmuth, *Inequitable Exposure to Air Pollution from Vehicles in California* (Cambridge, MA: Union of Concerned Scientists, 2019), 201, https://www.ucsusa.org/resources/inequitable-exposure-air-pollution-vehicles-california-2019.

14. Douglas S. Massey and Nancy A. Denton, "American Apartheid: Segregation and the Making of the Underclass," in *Social Stratification, Class, Race, and Gender in Sociological Perspective*, ed. David B. Grusky (Oxfordshire: Routledge, 2001), 660–670.

15. D. Ruiz, M. Becerra, J. S. Jagai, K. Ard, and R. M. Sargis, "Disparities in Environmental Exposures to Endocrine-Disrupting Chemicals and Diabetes Risk in Vulnerable Populations," *Diabetes Care* 41, no. 1 (2018): 193–205, 201.

16. The Ella Roberta Foundation, "Clean Air for All," https://ellaroberta.org.

17. Kimberly Warner, November 8, 2022, Zoom interview with the authors.

18. Warner, interview, November 8, 2022.

19. Warner, interview, November 8, 2022.

20. Health Effects Institute, *State of Global Air 2020*, Special Report (Boston: Health Effects Institute, 2020), https://www.stateofglobalair.org.

21. Arshad R. Zargar, "Toxic Smog Turns India's Capital 'into a Gas Chamber,'" *CBS News*, November 4, 2022.

22. Reuters, "Delhi's Air Branded 'Hazardous,' Spurs Calls to Close Schools," *Reuters*, November 3, 2022, https://www.reuters.com/world/india/delhis-air-crime-against-humanity-spurs-calls-close-schools-2022-11-03.

23. Health Effects Institute, *State of Global Air 2020*.

24. William A. Suk, Hamid Ahanchian, Kwadwo Ansong Asante, David O. Carpenter, Fernando Diaz-Barriga, Eun-Hee Ha, Xia Huo, Malcolm King, et. al, "Environmental Pollution: An Under-Recognized Threat to Children's Health, Especially in Low- and Middle-Income Countries," *Environmental Health Perspectives* 124, no. 3 (2016): A41–A45.

25. Jin Wu, Derek Watkins, Josh Williams, Shalini Venugopal Bhagat, Hari Kumar, and Jeffrey Gettlemans, "Who Gets to Breathe Clean Air in New Delhi?," *New York Times*. December 17, 2020, https://www.nytimes.com/interactive/2020/12/17/world/asia/india-pollution-inequality.html.

CHAPTER 7

1. C. A. Pope III, M. Ezzati, and D. W. Dockery, "Fine-Particulate Air Pollution and Life Expectancy in the United States," *New England Journal of Medicine* 360 (2009): 376–386.

2. Stephen Colbert, "Cheating Death—Lung Health," *The Colbert Report*, January 27, 2009, https://www.cc.com/video/ei15xx/the-colbert-report-cheating-death-lung-health.

3. See additional documentation elsewhere: Douglas W. Dockery and C. Arden Pope, "Lost Life Expectancy Due to Air Pollution in China," *Risk Dialogue Magazine* 17 (2014): 2–48, https://www.swissre.com/dam/jcr:ca35b02a-dd17-49bf-99ef-076369cf7ff7/CGD_RDS_Health_Risk_Factors_Web.pdf.

4. Joshua S. Apte, Michael Brauer, Aaron J. Cohen, Majid Ezzati, and C. Arden Pope, "Ambient $PM_{2.5}$ Reduces Global and Regional Life Expectancy," *Environmental Science & Technology Letters* 5, no. 9 (2018): 546–551.

5. Dockery and Pope, "Lost Life Expectancy Due to Air Pollution in China."

6. Prabhat Jha, Chinthanie Ramasundarahettige, Victoria Landsman, Brian Rostron, Michael Thun, Robert N. Anderson, Tim McAfee, and Richard Peto, "Twenty-First-Century Hazards of Smoking and Benefits of Cessation in the United States," *New England Journal of Medicine* 368, no. 4 (2013): 341–350.

7. I. Kelly and L. Clancy, "Mortality in a General Hospital and Urban Air Pollution," *Irish Medical Journal* Oct. 77, no. 10 (1984): 322–324, PMID: 6500894.

8. Luke Clancy, Pat Goodman, Hamish Sinclair, and Douglas W Dockery, "Effect of Air-Pollution Control on Death Rates in Dublin, Ireland: An Intervention Study," *Lancet* 360, no. 9341 (2002): 1210–1214.

9. Patrick G. Goodman, David Q. Rich, Ariana Zeka, Luke Clancy, and Douglas W. Dockery, "Effect of Air Pollution Controls on Black Smoke and Sulfur Dioxide Concentrations across Ireland," *Journal of the Air & Waste Management Association* 59, no. 2 (2009): 207–213.

10. David Rich, Prethibha George, Patrick Goodman, Pamela Ohman-Strickland, Luke Clancy, Tanya Kotlov, and Douglas Dockery, "Effect of Air Pollution Control on Mortality in County Cork, Ireland," *Epidemiology* 20 (2009), https://doi.org/10.1097/01.ede.0000362914.17454.32.

11. John Trijonis, "Visibility in the Southwest: An Exploration of the Historical Database (1967)," *Atmospheric Environment* 13, no. 6 (1979): 833–843.

12. John Trijonis, "Vegas Winners," *Engineering & Science* 2 (2001): 35–41, https://calteches.library.caltech.edu/682/2/Vegas.pdf.

13. Trijonis, "Vegas Winners," 35.

14. Trijonis, "Vegas Winners," 36.

15. Trijonis, "Vegas Winners," 36–37.

16. Trijonis, "Vegas Winners," 41.

17. Trijonis, "Vegas Winners," 41.

18. C. Arden Pope, Douglas L. Rodermund, and Matthew M. Gee, "Mortality Effects of a Copper Smelter Strike and Reduced Ambient Sulfate Particulate Matter Air Pollution," *Environmental Health Perspectives* 115, no. 5 (2007): 679–683.

19. Majid Ezzati, Ari B. Friedman, Sandeep C. Kulkarni, and Christopher J. Murray, "The Reversal of Fortunes: Trends in County Mortality and Cross-County Mortality Disparities in the United States," *PLoS Medicine* 5, no. 4 (2008): 0557–0568.

20. Pope, Ezzati, and Dockery, "Fine-Particulate Air Pollution and Life Expectancy."

21. Goran Krstić, "A Reanalysis of Fine Particulate Matter Air Pollution versus Life Expectancy in the United States," *Journal of the Air & Waste Management Association* 63, no. 2 (2013): 133–135.

22. C. Arden Pope, Majid Ezzati, and Douglas. W. Dockery, "Fine Particulate Air Pollution and Life Expectancies in the United States: The Role of Influential Observations," *Journal of the Air & Waste Management Association* 63, no. 2 (2013): 129–132.

23. S. Y. Kim, C. A. Pope III, J. D. Marshall, N. Fann, and L. Sheppard, "Reanalysis of the Association between Reduction in Long-Term $PM_{2.5}$ Concentrations and Improved Life Expectancy," *Environmental Health* 20, no. 1 (2021): 1–10.

24. A. W. Correia, C. A. Pope III, D. W. Dockery, Y. Wang, M. Ezzati, and F. Dominici, "Effect of Air Pollution Control on Life Expectancy in the United States: An

Analysis of 545 U.S. Counties for the Period from 2000 to 2007," *Epidemiology* 24, no. 1 (2013): 23–31; Francesca Dominici, Yun Wang, Andrew W. Correia, Majid Ezzati, C. Arden Pope, and Douglas W. Dockery, "Chemical Composition of Fine Particulate Matter and Life Expectancy," *Epidemiology* 26, no. 4 (2015): 556–564.

25. Francesca Dominici, Michael Greenstone, and Cass R. Sunstein, "Particulate Matter Matters," *Science* 344, no. 6181 (2014): 257–259.

26. K. Y. Chay and M. Greenstone, "The Impact of Air Pollution on Infant Mortality: Evidence from Geographic Variation in Pollution Shocks Induced by a Recession," *Quarterly Journal of Economics* 118, no. 3 (2003): 1121–1167.

27. Avraham Ebenstein, Maoyong Fan, Michael Greenstone, Guojun He, and Maigeng Zhou, "New Evidence on the Impact of Sustained Exposure to Air Pollution on Life Expectancy from China's Huai River Policy," *Proceedings of the National Academy of Sciences* 114, no. 29 (2017): 10384–10389.

28. UChicago News, "Air Pollution Reduces Global Life Expectancy by Nearly Two Years," November 19, 2018, https://news.uchicago.edu/story/air-pollution-reduces -global-life-expectancy-nearly-two-years.

29. W. James Gauderman, Edward Avol, Frank Gilliland, Hita Vora, Duncan Thomas, Kiros Berhane, Rob McConnell, et al., "The Effect of Air Pollution on Lung Development from 10 to 18 Years of Age," *New England Journal of Medicine* 351, no. 11 (2004): 1057–1067.

30. W. James Gauderman, Hita Vora, Rob McConnell, Kiros Berhane, Frank Gilliland, Duncan Thomas, Fred Lurmann, et al., "Effect of Exposure to Traffic on Lung Development from 10 to 18 Years of Age: A Cohort Study," *Lancet* 369, no. 9561 (2007): 571–577.

31. W. James Gauderman, Robert Urman, Edward Avol, Kiros Berhane, Rob McConnell, Edward Rappaport, Roger Chang, Fred Lurmann, and Frank Gilliland, "Association of Improved Air Quality with Lung Development in Children," *New England Journal of Medicine* 372, no. 10 (2015): 905–913, 905.

32. Douglas W. Dockery and James H. Ware, "Cleaner Air, Bigger Lungs," *New England Journal of Medicine* 372, no. 10 (2015): 970–972.

33. Kiros Berhane, Chih-Chieh Chang, Rob McConnell, W. James Gauderman, Edward Avol, Ed Rappaport, Robert Urman, Fred Lurmann, and Frank Gilliland, "Association of Changes in Air Quality with Bronchitic Symptoms in Children in California, 1993–2012," *JAMA* 315, no. 14 (2016): 1491–1501.

34. Erika Garcia, Kiros T. Berhane, Talat Islam, Rob McConnell, Robert Urman, Zhanghua Chen, and Frank D. Gilliland, "Association of Changes in Air Quality with Incident Asthma in Children in California, 1993–2014," *JAMA* 321, no. 19 (2019): 1906–1915.

35. Edward L. Avol, James Gauderman, Sylvia M. Tan, Stephanie J. London, and John M. Peters, "Respiratory Effects of Relocating to Areas of Differing Air Pollution Levels," *American Journal of Respiratory and Critical Care Medicine* 164, no. 11 (2001): 2067–2072.

36. Hong Chen, Jay S. Kaufman, Toyib Olaniyan, Lauren Pinault, Michael Tjep-kema, Li Chen, Aaron van Donkelaar, et al., "Changes in Exposure to Ambient Fine Particulate Matter after Relocating and Long-Term Survival in Canada: Quasi-Experimental Study," *BMJ* 375, no. 2368 (2021), https://doi.org/10.1136/bmj.n2368.

37. Gavin Pereira, "Cut Particulate Air Pollution, Save Lives," *BMJ* 275, no. 2561 (2021), https://doi.org/10.1136/bmj.n2561.

38. Francine Laden, Joel Schwartz, Frank E. Speizer, and Douglas W. Dockery, "Reduction in Fine Particulate Air Pollution and Mortality," *American Journal of Respiratory and Critical Care Medicine* 173, no. 6 (2006): 667–672; Johanna Lepeule, Francine Laden, Douglas Dockery, and Joel Schwartz, "Chronic Exposure to Fine Particles and Mortality: An Extended Follow-Up of the Harvard Six Cities Study from 1974 to 2009," *Environmental Health Perspectives* 120, no. 7 (2012): 965–970.

39. Trijonis, "Vegas Winners," 41.

CHAPTER 8

1. Marlo Lewis Jr., "CEI Comments on EPA's Proposed Review of the National Ambient Air Quality Standards for Particulate Matter," *Competitive Enterprise Institute*, Docket ID No. EPA-HQ-OAR-2015-0072, June 30, 2020, https://cei.org/regulatory_comments/cei-comments-on-epas-proposed-review-of-the-national-ambient-air-quality-standards-for-particulate-matter/#_ftn7.

2. Herbert L. Needleman, Charles Gunnoe, Alan Leviton, Robert Reed, Henry Peresie, Cornelius Maher, and Peter Barrett, "Deficits in Psychologic and Classroom Performance of Children with Elevated Dentine Lead Levels," *New England Journal of Medicine* 300, no. 13 (1979): 689–695.

3. Richard L. Canfield, Charles R. Henderson, Deborah A. Cory-Slechta, Christopher Cox, Todd A. Jusko, and Bruce P. Lanphear, "Intellectual Impairment in Children with Blood Lead Concentrations below 10 mg per Deciliter," *New England Journal of Medicine* 348, no. 16 (2003): 1517–1526.

4. B. E. Brown, "The Environmental Protection Agency's Research Program with Primary Emphasis on the Community Health and Environmental Surveillance System (CHESS): An Investigative Report" (Washington, DC: U.S. Government Printing Office, 1976).

5. James E. Enstrom, "Scientific Misconduct in Fine Particulate Matter Epidemiology by Dr. C. Arden Pope, III, in Collaboration with Drs. Daniel Krewski, Michael Jerrett, and Richard Burnett, with the Complete Cooperation of the American Cancer Society," Scientific Integrity Institute, November 15, 2013, http://www.scientificintegrity institute.org/pope111513.pdf.

6. Steven Milloy, letter to Dr. Kathy Partin, director of Office of Research Integrity, U.S. Department of Health and Human Services, Request for Investigation of Research Misconduct, JunkScience.com, September 5, 2017, https://junkscience .com/wp-content/uploads/2017/09/Milloy-RFI-to-ORI-09052017-Final.pdf.

7. *Wall Street Journal*, "NIH Finds No Fraud in Air Quality Study," December 7, 1990.

8. Carolyn Kormann, "Scott Pruitt's Crusade against 'Secret Science' Could Be Disastrous for Public Health," *New Yorker*, April 26, 2018, para. 8, https://www.new yorker.com/science/elements/scott-pruitts-crusade-against-secret-science-could-be -disastrous-for-public-health.

9. Steven J. Milloy, "Pope-a-Dope?," November 8, 1997, http://junksciencearchive .com/news/ali.html.

10. C. Arden Pope, Michael J. Thun, Mohan M. Namboodiri, Douglas W. Dockery, John S. Evans, Frank E. Speizer, and Clark W. Heath, "Particulate Air Pollution as a Predictor of Mortality in a Prospective Study of U.S. Adults," *American Journal of Respiratory and Critical Care Medicine* 151, no. 3.1 (1995): 669–674.

11. Working Group on Public Health and Fossil Fuel Combustion, "Short-Term Improvements in Public Health from Global-Climate Policies on Fossil-Fuel Combustion: An Interim Report," *Lancet* 350, no. 9088 (1997): 1341–1349.

12. C. Arden Pope, Richard T. Burnett, Daniel Krewski, Michael Jerrett, Yuanli Shi, Eugenia E. Calle, and Michael J. Thun, "Cardiovascular Mortality and Exposure to Airborne Fine Particulate Matter and Cigarette Smoke," *Circulation* 120, no. 11 (2009): 941–948; C. Arden Pope, Richard T. Burnett, Michelle C. Turner, Aaron Cohen, Daniel Krewski, Michael Jerrett, Susan M. Gapstur, and Michael J. Thun, "Lung Cancer and Cardiovascular Disease Mortality Associated with Ambient Air Pollution and Cigarette Smoke: Shape of the Exposure–Response Relationships," *Environmental Health Perspectives* 119, no. 11 (2011): 1616–1621.

13. Pope, Burnett, Krewski, et al., "Cardiovascular Mortality and Exposure to Airborne Fine Particulate Matter and Cigarette Smoke."

14. Pope, Burnett, Turner, et al., "Lung Cancer and Cardiovascular Disease Mortality Associated with Ambient Air Pollution and Cigarette Smoke."

15. Richard Burnett, Hong Chen, Mieczysław Szyszkowicz, Neal Fann, Bryan Hubbell, C. Arden Pope, Joshua S. Apte, et al., "Global Estimates of Mortality Associated with Long-Term Exposure to Outdoor Fine Particulate Matter," *Proceedings of the National Academy of Sciences* 115, no. 38 (2018): 9592–9597; Aaron J. Cohen, Michael Brauer, Richard Burnett, H. Ross Anderson, Joseph Frostad, Kara Estep, Kalpana Balakrishnan, et al., "Estimates and 25-Year Trends of the Global Burden of Disease Attributable to Ambient Air Pollution: An Analysis of Data from the Global Burden of Diseases Study 2015," *Lancet* 389, no. 10082 (2017): 1907–1918.

16. Pope, Burnett, Turner, et al., "Shape of the Exposure–Response Relationships"; Allan Hackshaw, Joan K. Morris, Sadie Boniface, Jin-Ling Tang, and Dušan Milenković, "Low Cigarette Consumption and Risk of Coronary Heart Disease and Stroke: Meta-Analysis of 141 Cohort Studies in 55 Study Reports," *BMJ* 360, no. 5855 (2018), https://doi.org/10.1136/bmj.j5855; Office on Smoking and Health (U.S.), *The Health Consequences of Involuntary Exposure to Tobacco Smoke: A Report of the Surgeon General* (Atlanta: Centers for Disease Control and Prevention, 2006), https:// www.ncbi.nlm.nih.gov/books/NBK44324; Koon K. Teo, Stephanie Ounpuu, Steven Hawken, M. R. Pandey, Vicent Valentin, David Hunt, Rafael Diaz, et al., "Tobacco Use and Risk of Myocardial Infarction in 52 Countries in the INTERHEART Study:

A Case-Control Study," *Lancet* 368, no. 9536 (2006): 647–658; Florian Fischer and Alexander Kraemer, "Meta-Analysis of the Association between Second-Hand Smoke Exposure and Ischaemic Heart Diseases, COPD, and Stroke," *BMC Public Health* 15, no. 1 (2015), https://doi.org/10.1186/s12889-015-2489-4; Jie Chen and Gerard Hoek, "Long-Term Exposure to PM and All-Cause and Cause-Specific Mortality: A Systematic Review and Meta-Analysis," *Environment International* 143 (2020): 105974; C. Arden Pope, Aaron J. Cohen, and Richard T. Burnett, "Cardiovascular Disease and Fine Particulate Matter," *Circulation Research* 122, no. 12 (2018): 1645–1647.

17. Richard T. Burnett, C. Arden Pope, Majid Ezzati, Casey Olives, Stephen S. Lim, Sumi Mehta, Hwashin H. Shin, et al., "An Integrated Risk Function for Estimating the Global Burden of Disease Attributable to Ambient Fine Particulate Matter Exposure," *Environmental Health Perspectives* 122, no. 4 (2014): 397–403.

18. Burnett, Chen, Szyszkowicz, et al., "Global Estimates of Mortality."

19. Richard T. Burnett, Joseph V. Spadaro, George R. Garcia, and C. Arden Pope, "Designing Health Impact Functions to Assess Marginal Changes in Outdoor Fine Particulate Matter," *Environmental Research* 204 (2022): 112245.

20. U.S. Food and Drug Administration, "Harmful and Potentially Harmful Constituents in Tobacco Products and Tobacco Smoke: Established List," April 2012.

CHAPTER 9

1. Jason Plautz, "Trump's Air Pollution Advisor: No Proof Cleaning Up Smog Saves Lives," *Reveal News*, October 24, 2018, https://revealnews.org/article/trumps-air-pollution-adviser-clean-air-saves-no-lives; Scott Waldman, "EPA Science Advisor Allowed Industry Group to Edit Journal Article," *Science*, December 10, 2018, https://www.science.org/content/article/epa-science-adviser-allowed-industry-group-edit-journal-article#:~:text=Cox%20is%20a%20statistician%20who,are%20breathing%20is%20hurting%20them.

2. Jeff Tollefson, "Air Pollution Science under Siege at U.S. Environment Agency," *Nature* 568, no. 7750 (2019): 15–16, https://doi.org/10.1038/d41586-019-00937-w.

3. Myron Ebell, "Harvard Junk Science Study Claims High Pollution Levels Increase Deaths from COVID-19," Competitive Enterprise Institute blog, May 11, 2020, https://cei.org/blog/harvard-junk-science-study-claims-high-pollution-levels-increase-deaths-from-covid-19.

4. Austin Bradford Hill, "The Environment and Disease: Association or Causation?," *Proceedings of the Royal Society of Medicine* 58, no. 5 (1965): 295–300; J. D. Angrist and J. S. Pischke, *Mostly Harmless Econometrics: An Empiricist's Companion* (Princeton, NJ: Princeton University Press, 2008); Michael Greenstone and Ted Gayer, "Quasi-Experimental and Experimental Approaches to Environmental Economics," *Journal of Environmental Economics and Management* 57, no. 1 (2009): 21–44; J. Peters, D. Janzing, and B. Schoelkopf, *Elements of Causal Inference: Foundations and Learning Algorithms* (Cambridge, MA: MIT Press, 2017); Francesca Dominici, Falco J. Bargagli-Stoffi, and Fabrizia Mealli, "From Controlled to Undisciplined Data: Estimating

Causal Effects in the Era of Data Science Using a Potential Outcome Framework," *Harvard Data Science Review* 3, no. 3 (2021), https://doi.org/10.1162/99608f92.8102afed.

5. Louis Anthony Cox, "Modernizing the Bradford Hill Criteria for Assessing Causal Relationships in Observational Data," *Critical Reviews in Toxicology* 48, no. 8 (2018): 682–712.

6. Louis Anthony Cox, "A Causal Analytics Toolkit (CAT) for Assessing Potential Causal Relations in Data," Regulatory Studies Center, George Washington University, May 11, 2016.

7. Gretchen T. Goldman and Francesca Dominici, "Don't Abandon Evidence and Process on Air Pollution Policy," *Science* 363, no. 6434 (2019): 1398–1400.

8. Francesca Dominici, Michael Greenstone, and Cass R. Sunstein, "Particulate Matter Matters," *Science* 344, no. 6181 (2014): 257–259; Corwin M. Zigler, Francesca Dominici, and Yun Wang, "Estimating Causal Effects of Air Quality Regulations Using Principal Stratification for Spatially Correlated Multivariate Intermediate Outcomes," *Biostatistics* 13, no. 2 (2012): 289–302; Corwin M. Zigler, Chanmin Kim, Christine Choirat, John Barrett Hansen, Yun Wang, Lauren Hund, Jonathan Samet, Gary King, Francesca Dominici, and HEI Health Review Committee, "Causal Inference Methods for Estimating Long-Term Health Effects of Air Quality Regulations," *Research Reports, Health Effects Institute* 187 (2016): 5–49; Francesca Dominici and Corwin Zigler, "Best Practices for Gauging Evidence of Causality in Air Pollution Epidemiology," *American Journal of Epidemiology* 186, no. 12 (2017): 1303–1309; Joel D. Schwartz, Yan Wang, Itai Kloog, Ma'ayan Yitshak-Sade, Francesca Dominici, and Antonella Zanobetti, "Estimating the Effects of $PM_{2.5}$ on Life Expectancy Using Causal Modeling Methods," *Environmental Health Perspectives* 126, no. 12 (2018): 127002.

9. Hill, "The Environment and Disease."

10. Hill, "The Environment and Disease," 299.

11. *Federal Register*, "Revisions to the National Ambient Air Quality Standards for Particulate Matter," *Federal Register* 52, no. 126, July 1, 1987, Rules and Regulations 24634 Environmental Protection Agency 40 CFR Part 50 [AD-FRL 3141–9(a)].

12. Douglas W. Dockery, James H. Ware, Benjamin G. Ferris, Frank E. Speizer, Nancy R. Cook, and Stanislaw M. Herman, "Change in Pulmonary Function in Children Associated with Air Pollution Episodes," *Journal of the Air Pollution Control Association* 32, no. 9 (1982): 937–942.

13. W. B. Dassen, B. Brunekreef, G. Hoek, P. Hofschreuder, B. Staatsen, H. de Groot, E. Schouten, and K. Biersteker, "Decline in Children's Pulmonary Function during an Air Pollution Episode," *Journal of the Air Pollution Control Association* 36, no. 11 (1986): 1223–1227.

14. Recent Integrated Science Assessment documents include U.S. EPA, Integrated Science Assessment (ISA) for Particulate Matter (Final Report, Dec. 2009), U.S. Environmental Protection Agency, Washington, DC, EPA/600/R-08/139F; U.S. EPA,

Integrated Science Assessment (ISA) for Particulate Matter (Final Report, Dec. 2019), U.S. Environmental Protection Agency, Washington, DC, EPA/600/R-19/188.

15. Jocelyn Kaiser, "Showdown over Clean Air Science," *Science* 277, no. 5325 (1997): 466–469. See also John Carey, "Tiny Particles, Big Dilemma: The Air Pollution Data Aren't Clear—So Should the EPA Be Making Rules?," *Business Week*, August 4, 1997, https://www.bloomberg.com/news/articles/1997-08-03/tiny-particles-big -dilemma#xj4y7vzkg.

16. C. Arden Pope, "Respiratory Disease Associated with Community Air Pollution and a Steel Mill, Utah Valley," *American Journal of Public Health* 79, no. 5 (1989): 623–628.

17. Joel Schwartz and Douglas W. Dockery, "Increased Mortality in Philadelphia Associated with Daily Air Pollution Concentrations," *American Review of Respiratory Disease* 145, no. 3 (1992): 600–604.

18. Douglas W. Dockery, C. Arden Pope, Xiping Xu, John D. Spengler, James H. Ware, Martha E. Fay, Benjamin G. Ferris, and Frank E. Speizer, "An Association between Air Pollution and Mortality in Six U.S. Cities," *New England Journal of Medicine* 329, no. 24 (1993): 1753–1759; C. Arden Pope, Michael J. Thun, Mohan M. Namboodiri, Douglas W. Dockery, John S. Evans, Frank E. Speizer, and Clark W. Heath, "Particulate Air Pollution as a Predictor of Mortality in a Prospective Study of U.S. Adults," *American Journal of Respiratory and Critical Care Medicine* 151, no. 3.1 (1995): 669–674.

19. David V. Bates, "Health Indices of the Adverse Effects of Air Pollution: The Question of Coherence," *Environmental Research* 59, no. 2 (1992): 336–349.

20. Douglas W. Dockery and C. Arden Pope, "Acute Respiratory Effects of Particulate Air Pollution," *Annual Review of Public Health* 15 (1994): 107–132; C. Arden Pope, D. V. Bates, and M. E. Raizenne, "Health Effects of Particulate Air Pollution: Time for Reassessment?," *Environmental Health Perspectives* 103 (1995): 472–480; C. Arden Pope and Douglas W. Dockery, "Health Effects of Fine Particulate Air Pollution: Lines That Connect," *Journal of the Air & Waste Management Association* 56, no. 6 (2006): 709–742; Robert D. Brook, Sanjay Rajagopalan, C. Arden Pope, Jeffrey R. Brook, Aruni Bhatnagar, Ana V. Diez-Roux, Fernando Holguin, et al., "Particulate Matter Air Pollution and Cardiovascular Disease," *Circulation* 121, no. 21 (2010): 2331–2378.

21. Michelle C. Turner, Zorana J. Andersen, Andrea Baccarelli, W. Ryan Diver, Susan M. Gapstur, C. Arden Pope, Diddier Prada, Jonathan Samet, George Thurston, and Aaron Cohen, "Outdoor Air Pollution and Cancer: An Overview of the Current Evidence and Public Health Recommendations," *CA: A Cancer Journal for Clinicians* 70, no. 6 (2020): 460–479.

22. C. Arden Pope III, R. T. Burnett, M. J. Thun, E. E. Calle, D. Krewski, K. Ito, and G. D. Thurston, "Lung Cancer, Cardiopulmonary Mortality, and Long-Term Exposure to Fine Particulate Air Pollution." *JAMA* 287, no. 9 (2002): 1132–1141; D. Krewski, M. Jerrett, R. T. Burnett, R. Ma, E. Hughes, Y. Shi, M. C. Turner, C. A. Pope III, G. Thurston, E. E. Calle, and M. J. Thun, *Extended Follow-Up and Spatial*

Analysis of the American Cancer Society Study Linking Particulate Air Pollution and Mortality (Boston: Health Effects Institute, 2009); Michael Jerrett, Michelle C. Turner, Bernardo S. Beckerman, C. Arden Pope, Aaron van Donkelaar, Randall V. Martin, Marc Serre, et al., "Comparing the Health Effects of Ambient Particulate Matter Estimated Using Ground-Based versus Remote Sensing Exposure Estimates," *Environmental Health Perspectives* 125, no. 4 (2017): 552–559.

23. C. Arden Pope and Richard T. Burnett, "Confounding in Air Pollution Epidemiology," *Epidemiology* 18, no. 4 (2007): 424–426.

24. Pope and Dockery, "Health Effects of Fine Particulate Air Pollution: Lines That Connect."

25. C. Arden Pope, "Respiratory Disease Associated with Community Air Pollution and a Steel Mill, Utah Valley," *American Journal of Public Health* 79, no. 5 (1989): 623–628; C. Arden Pope, "Respiratory Hospital Admissions Associated with PM_{10} Pollution in Utah, Salt Lake, and Cache Valleys," *Archives of Environmental Health: An International Journal* 46, no. 2 (1991): 90–97; Michael R. Ransom and C. Arden Pope, "Elementary School Absences and PM_{10} Pollution in Utah Valley," *Environmental Research* 58, nos. 1–2 (1992): 204–219.

26. Robert B. Penfold and Fang Zhang, "Use of Interrupted Time Series Analysis in Evaluating Health Care Quality Improvements," *Academic Pediatrics* 13, no. 6 (2013): S38–S44.

27. Luke Clancy, Pat Goodman, Hamish Sinclair, and Douglas W Dockery, "Effect of Air-Pollution Control on Death Rates in Dublin, Ireland: An Intervention Study," *Lancet* 360, no. 9341 (2002): 1210–1214.

28. C. Arden Pope, Douglas L. Rodermund, and Matthew M. Gee, "Mortality Effects of a Copper Smelter Strike and Reduced Ambient Sulfate Particulate Matter Air Pollution," *Environmental Health Perspectives* 115, no. 5 (2007): 679–683.

29. Guido W. Imbens and Thomas Lemieux, "Regression Discontinuity Designs: A Guide to Practice," *Journal of Econometrics* 142, no. 2 (2008): 615–635.

30. Avraham Ebenstein, Maoyong Fan, Michael Greenstone, Guojun He, and Maigeng Zhou, "New Evidence on the Impact of Sustained Exposure to Air Pollution on Life Expectancy from China's Huai River Policy," *Proceedings of the National Academy of Sciences* 114, no. 29 (2017): 10384–10389.

31. Angrist and Pischke, *Mostly Harmless Econometrics*.

32. C. A. Pope III, M. Ezzati, and D. W. Dockery, "Fine-Particulate Air Pollution and Life Expectancy in the United States," *New England Journal of Medicine* 360 (2009): 376–386.

33. Nicholas J. Sanders, Alan I. Barreca, and Matthew J. Neidell, "Estimating Causal Effects of Particulate Matter Regulation on Mortality," *Epidemiology* 31, no. 2 (2020): 160–167.

34. Nicholas M. Hales, Caleb C. Barton, Michael R. Ransom, Ryan T. Allen, and C. Arden Pope, "A Quasi-Experimental Analysis of Elementary School Absences and

Fine Particulate Air Pollution," *Medicine* 95, no. 9 (2016), https://doi.org/10.1097/md.0000000000002916.

35. Sanders, Barreca, and Neidell, "Estimating Causal Effects."

36. Joel Schwartz, Marie-Abele Bind, and Petros Koutrakis, "Estimating Causal Effects of Local Air Pollution on Daily Deaths: Effect of Low Levels," *Environmental Health Perspectives* 125, no. 1 (2017): 23–29.

37. Joel Schwartz, Kelvin Fong, and Antonella Zanobetti, "A National Multicity Analysis of the Causal Effect of Local Pollution, NO_2, and $PM_{2.5}$ on Mortality," *Environmental Health Perspectives* 126, no. 8 (2018): 087004.

38. Tatyana Deryugina, Garth Heutel, Nolan H. Miller, David Molitor, and Julian Reif, "The Mortality and Medical Costs of Air Pollution: Evidence from Changes in Wind Direction," *American Economic Review* 109, no. 12 (2019): 4178–4219.

39. Peter C. Austin, "An Introduction to Propensity Score Methods for Reducing the Effects of Confounding in Observational Studies," *Multivariate Behavioral Research* 46, no. 3 (2011): 399–424.

40. Hong Chen, Jay S. Kaufman, Toyib Olaniyan, Lauren Pinault, Michael Tjepkema, Li Chen, Aaron van Donkelaar, et al., "Changes in Exposure to Ambient Fine Particulate Matter after Relocating and Long-Term Survival in Canada: Quasi-Experimental Study," *BMJ* 375, no. 2368 (2021), https://doi.org/10.1136/bmj.n2368.

41. Yaguang Wei, Yan Wang, Xiao Wu, Qian Di, Liuhua Shi, Petros Koutrakis, Antonella Zanobetti, Francesca Dominici, and Joel D. Schwartz, "Causal Effects of Air Pollution on Mortality Rate in Massachusetts," *American Journal of Epidemiology* 189, no. 11 (2020): 1316–1323.

42. Yan Wang, Mihye Lee, Pengfei Liu, Liuhua Shi, Zhi Yu, Yara Abu Awad, Antonella Zanobetti, and Joel D. Schwartz, "Doubly Robust Additive Hazards Models to Estimate Effects of a Continuous Exposure on Survival," *Epidemiology* 28, no. 6 (2017): 771–779.

43. Schwartz, Fong, and Zanobetti, "A National Multicity Analysis of the Causal Effect."

44. Qian Di, Yan Wang, Antonella Zanobetti, Yun Wang, Petros Koutrakis, Christine Choirat, Francesca Dominici, and Joel D. Schwartz, "Air Pollution and Mortality in the Medicare Population," *New England Journal of Medicine* 376, no. 26 (2017): 2513–2522.

45. X. Wu, D. Braun, J. Schwartz, M. A. Kioumourtzoglou, and F. Dominici, "Evaluating the Impact of Long-Term Exposure to Fine Particulate Matter on Mortality among the Elderly," *Science Advances* 6, no. 29 (2020), https://doi.org/10.1126/sciadv.aba5692.

46. Yaguang Wei, Mahdieh Danesh Yazdi, Qian Di, Weeberb J. Requia, Francesca Dominici, Antonella Zanobetti, and Joel Schwartz, "Emulating Causal Dose-Response Relations between Air Pollutants and Mortality in the Medicare Population," *Environmental Health* 20, no. 53 (2021), https://doi.org/10.1186/s12940-021-00742-x.

47. C. Arden Pope, Jacob S. Lefler, Majid Ezzati, Joshua D. Higbee, Julian D. Marshall, Sun-Young Kim, Matthew Bechle, et al., "Mortality Risk and Fine Particulate Air Pollution in a Large, Representative Cohort of U.S. Adults," *Environmental Health Perspectives* 127, no. 7 (2019): 077007; Jacob S. Lefler, Joshua D. Higbee, Richard T. Burnett, Majid Ezzati, Nathan C. Coleman, Dalton D. Mann, Julian D. Marshall, et al., "Air Pollution and Mortality in a Large, Representative U.S. Cohort: Multiple-Pollutant Analyses, and Spatial and Temporal Decompositions," *Environmental Health* 18, no. 1 (2019), https://doi.org/10.1186/s12940-019-0544-9; Nathan C. Coleman, Richard T. Burnett, Joshua D. Higbee, Jacob S. Lefler, Ray M. Merrill, Majid Ezzati, Julian D. Marshall, et al., "Cancer Mortality Risk, Fine Particulate Air Pollution, and Smoking in a Large, Representative Cohort of U.S Adults," *Cancer Causes & Control* 31, no. 8 (2020): 767–776.

48. C. Arden Pope, Jacob S. Lefler, Majid Ezzati, Joshua D. Higbee, Julian D. Marshall, Sun-Young Kim, Matthew Bechle, et al., "Mortality Risk and Fine Particulate Air Pollution in a Large, Representative Cohort of U.S. Adults," *Environmental Health Perspectives* 127, no. 7 (2019): 077007.

49. J. D. Higbee, J. S. Lefler, R. T. Burnett, M. Ezzati, J. D. Marshall, S. Y. Kim, M. Bechle, A. L. Robinson, and C. A. Pope III, "Estimating Long-Term Pollution Exposure Effects through Inverse Probability Weighting Methods with Cox Proportional Hazards Models," *Environmental Epidemiology* 4, no. 2 (2020): e085, 1–10.

CHAPTER 10

1. W. J. Madia, "A Call for More Science in EPA Regulations," *Science* 282, no. 5386 (1998): 45, para. 1.

2. Sverre Vedal, "Ambient Particles and Health: Lines That Divide," *Journal of the Air & Waste Management Association* 47, no. 5 (1997): 551–581, 552 and 558.

3. C. Arden Pope and Douglas W. Dockery, "Health Effects of Fine Particulate Air Pollution: Lines That Connect," *Journal of the Air & Waste Management Association* 56, no. 6 (2006): 709–742.

4. Connie C. Hsia, Dallas M. Hyde, Matthias Ochs, and Ewald R. Weibel, "An Official Research Policy Statement of the American Thoracic Society/European Respiratory Society: Standards for Quantitative Assessment of Lung Structure," *American Journal of Respiratory and Critical Care Medicine* 181, no. 4 (2010): 394–418.

5. Peter Gehr and Joachim Heyder, *Particle-Lung Interactions* (New York: Marcel Dekker, 2000).

6. Andrew J. Ghio, "Biological Effects of Utah Valley Ambient Air Particles in Humans: A Review," *Journal of Aerosol Medicine* 17, no. 2 (2004): 157–164.

7. Douglas W. Dockery, James H. Ware, Benjamin G. Ferris, Frank E. Speizer, Nancy R. Cook, and Stanislaw M. Herman, "Change in Pulmonary Function in Children Associated with Air Pollution Episodes," *Journal of the Air Pollution Control Association* 32, no. 9 (1982): 937–942; Douglas W. Dockery, Frank E. Speizer, Daniel O. Stram, James H. Ware, John D. Spengler, and Benjamin G. Ferris, "Effects

of Inhalable Particles on Respiratory Health of Children," *American Review of Respiratory Disease* 139, no. 3 (1989): 587–594; B. Brunekreef, P. L. Kinney, J. H. Ware, D. Dockery, F. E. Speizer, J. D. Spengler, and B. G. Ferris, "Sensitive Subgroups and Normal Variation in Pulmonary Function Response to Air Pollution Episodes," *Environmental Health Perspectives*, no. 90 (1991): 189–193; Douglas W. Dockery, C. Arden Pope III, and Frank E. Speizer, "Effects of Particulate Air Pollution Exposures," in *Particle-Lung Interactions*, ed. Peter Gehr and Joachim Heyder, 671–703 (New York: Marcel Dekker, 2000).

8. Dockery, Ware, Ferris, et al., "Change in Pulmonary Function in Children."

9. Dockery, Ware, Ferris, et al., "Change in Pulmonary Function in Children."

10. C. Arden Pope, Douglas W. Dockery, John D. Spengler and Mark E. Raizenne, "Respiratory Health and PM_{10} Pollution: A Daily Time Series Analysis," *American Review of Respiratory Disease* 144, no. 3.1 (1991): 668–674; C. Arden Pope and Douglas W. Dockery, "Acute Health Effects of PM_{10} Pollution on Symptomatic and Asymptomatic Children," *American Review of Respiratory Disease* 145, no. 5 (1992): 1123–1128.

11. Lucas M. Neas, Douglas W. Dockery, Petros Koutrakis, David J. Tollerud, and Frank E. Speizer, "The Association of Ambient Air Pollution with Twice Daily Peak Expiratory Flow Rate Measurements in Children," *American Journal of Epidemiology* 141, no. 2 (1995): 111–122.

12. D. W. Dockery, J. Cunningham, A. I. Damokosh, L. M. Neas, J. D. Spengler, P. Koutrakis, J. H. Ware, M. Raizenne, and F. E. Speizer, "Health Effects of Acid Aerosols on North American Children: Respiratory Symptoms," *Environmental Health Perspectives* 104, no. 5 (1996): 500–505; M. Raizenne, L. M. Neas, A. I. Damokosh, D. W. Dockery, J. D. Spengler, P. Koutrakis, J. H. Ware, and F. E. Speizer, "Health Effects of Acid Aerosols on North American Children: Pulmonary Function," *Environmental Health Perspectives* 104, no. 5 (1996): 506–514.

13. Joel Schwartz, "Lung Function and Chronic Exposure to Air Pollution: A Cross-Sectional Analysis of NHANES II," *Environmental Research* 50, no. 2 (1989): 309–321; Joel Schwartz, "Particulate Air Pollution and Chronic Respiratory Disease," *Environmental Research* 62, no. 1 (1993): 7–13.

14. W. B. Dassen, B. Brunekreef, G. Hoek, P. Hofschreuder, B. Staatsen, H. de Groot, E. Schouten, and K. Biersteker, "Decline in Children's Pulmonary Function during an Air Pollution Episode," *Journal of the Air Pollution Control Association* 36, no. 11 (1986): 1223–1227; Willem Roemer, Gerard Hoek, and Bert Brunekreef, "Effect of Ambient Winter Air Pollution on Respiratory Health of Children with Chronic Respiratory Symptoms," *American Review of Respiratory Disease* 147, no. 1 (1993): 118–124; G. Hoek and B. Brunekreef, "Acute Effects of a Winter Air Pollution Episode on Pulmonary Function and Respiratory Symptoms of Children," *Archives of Environmental Health: An International Journal* 48, no. 5 (1993): 328–335; G. Hoek and B. Brunekreef, "Effects of Low-Level Winter Air Pollution Concentrations on Respiratory Health of Dutch Children,." *Environmental Research* 64, no. 2 (1994): 136–150.

15. W. James Gauderman, Edward Avol, Frank Gilliland, Hita Vora, Duncan Thomas, Kiros Berhane, Rob McConnell, et al., "The Effect of Air Pollution on Lung Development from 10 to 18 Years of Age," *New England Journal of Medicine* 351, no. 11 (2004): 1057–1067; W. James Gauderman, Hita Vora, Rob McConnell, Kiros Berhane, Frank Gilliland, Duncan Thomas, Fred Lurmann, et al., "Effect of Exposure to Traffic on Lung Development from 10 to 18 Years of Age: A Cohort Study," *Lancet* 369, no. 9561 (2007): 571–577; W. James Gauderman, Robert Urman, Edward Avol, Kiros Berhane, Rob McConnell, Edward Rappaport, Roger Chang, Fred Lurmann, and Frank Gilliland, "Association of Improved Air Quality with Lung Development in Children," *New England Journal of Medicine* 372, no. 10 (2015): 905–913; Kiros Berhane, Chih-Chieh Chang, Rob McConnell, W. James Gauderman, Edward Avol, Ed Rappaport, Robert Urman, Fred Lurmann, and Frank Gilliland, "Association of Changes in Air Quality with Bronchitic Symptoms in Children in California, 1993–2012," *JAMA* 315, no. 14 (2016): 1491–1501; Erika Garcia, Kiros T. Berhane, Talat Islam, Rob McConnell, Robert Urman, Zhanghua Chen, and Frank D. Gilliland, "Association of Changes in Air Quality with Incident Asthma in Children in California, 1993–2014," *JAMA* 321, no. 19 (2019): 1906–1915.

16. Benjamin D. Horne, Elizabeth A. Joy, Michelle G. Hofmann, Per H. Gesteland, John B. Cannon, Jacob S. Lefler, Denitza P. Blagev, et al., "Short-Term Elevation of Fine Particulate Matter Air Pollution and Acute Lower Respiratory Infection," *American Journal of Respiratory and Critical Care Medicine* 198, no. 6 (2018): 759–766.

17. Helmut Sies, *Oxidative Stress* (London: Academic Press, 1985); Helmut Sies, "Oxidative Stress: From Basic Research to Clinical Application," *American Journal of Medicine* 91, no. 3 (1991): 31S–38S.

18. Maura Lodovici and Elisabetta Bigagli, "Oxidative Stress and Air Pollution Exposure," *Journal of Toxicology* 201, no. 487074 (2011): 1–9.

19. F. J. Kelly, "Oxidative Stress: Its Role in Air Pollution and Adverse Health Effects," *Occupational and Environmental Medicine* 60, no. 8 (2003): 612–616.

20. I. S. Mudway, F. J. Kelly, and S. T. Holgate, "Oxidative Stress in Air Pollution Research," *Free Radical Biology and Medicine* 151 (2020): 2–6.

21. IARC, "Air Pollution and Cancer," IARC Scientific Publication No. 161 (Lyon: International Agency for Research on Cancer, 2013), https://www.iarc.who.int/news -events/iarc-scientific-publication-no-161-air-pollution-and-cancer-as-an-e-book.

22. Michelle C. Turner, Zorana J. Andersen, Andrea Baccarelli, W. Ryan Diver, Susan M. Gapstur, C. Arden Pope, Diddier Prada, Jonathan Samet, George Thurston, and Aaron Cohen, "Outdoor Air Pollution and Cancer: An Overview of the Current Evidence and Public Health Recommendations," *CA: A Cancer Journal for Clinicians* 70, no. 6 (2020): 460–479.

23. Joanna Strzelczyk and Andrzej Wiczkowski, "Oxidative Damage and Carcinogenesis," *Współczesna Onkologia* 3 (2012): 230–233, https://doi.org/10.5114/wo.2012 .29290.

24. P. M. Ridker, J. G. MacFadyen, T. Thuren, et al., "Effect of Interleukin-1β Inhibition with Canakinumab on Incident Lung Cancer in Patients with Atherosclerosis:

Exploratory Results from a Randomized, Double-Blind, Placebo-Controlled Trial," *Lancet* 390, no. 10105 (2017): 1833–1842.

25. Charles Swanton, "Mechanism of Action and an Actionable Inflammatory Axis for Air Pollution Induced Non-Small Cell Lung Cancer in Never Smokers," presented during Presidential Symposium at the ESMO Conference, September 10, 2022, *Annals of Oncology* 33, no. 7 (2022).

26. Elizabeth Gourd, "New Evidence That Air Pollution Contributes Substantially to Lung Cancer," *Lancet Oncology* 23, no. 10 (2022): w448, https://doi.org/10.1016/s1470-2045(22)00569-1.

27. Douglas W. Dockery, C. Arden Pope III, Richard E. Kanner, G. Martin Villegas, and Joel Schwartz, "Daily Changes in Oxygen Saturation and Pulse Rate Associated with Particulate Air Pollution and Barometric Pressure," Research Report No. 83, January 1999, Health Effects Institute.

28. John Godleski, Richard L. Verrier, Petro Koutrakis, and Paul Catalano, "Mechanisms of Morbidity and Mortality from Exposure to Ambient Air Particles," Research Report No. 91, February 2000, Health Effects Institute.

29. C. Arden Pope, Douglas W. Dockery, Richard E. Kanner, G. Martin Villegas, and Joel Schwartz, "Oxygen Saturation, Pulse Rate, and Particulate Air Pollution," *American Journal of Respiratory and Critical Care Medicine* 159, no. 2 (1999): 365–372.

30. C. Arden Pope, Richard L. Verrier, Eric G. Lovett, Andrew C. Larson, Mark E. Raizenne, Richard E. Kanner, Joel Schwartz, G. Martin Villegas, Diane R. Gold, and Douglas W. Dockery, "Heart Rate Variability Associated with Particulate Air Pollution," *American Heart Journal* 138, no. 5 (1995): 890–899.

31. Pope, Verrier, Lovett, et al., "Heart Rate Variability."

32. C. Arden Pope, Matthew L. Hansen, Russell W. Long, Karen R. Nielsen, Norman L. Eatough, William E. Wilson, and Delbert J. Eatough, "Ambient Particulate Air Pollution, Heart Rate Variability, and Blood Markers of Inflammation in a Panel of Elderly Subjects," *Environmental Health Perspectives* 112, no. 3 (2004): 339–345.

33. R. L. Lux and C. A. Pope III, "Air Pollution Effects on Ventricular Repolarization," *Health Effects Institute* 141 (2009), Research Report.

34. T. J. Bunch, B. D. Horne, S. J. Asirvatham, J. D. Day, B. G. Crandall, J. P. Weiss, J. S. Osborn, J. L. Anderson, J. B. Muhlestein, D. L. Lappe, and C. A. Pope III, "Atrial Fibrillation Hospitalization Is Not Increased with Short-Term Elevations in Exposure to Fine Particulate Air Pollution," *PACE: Pacing and Clinical Electrophysiology* 34 (2011): 1475–1479.

35. D. W. Dockery, H. Luttmann-Gibson, D. Q. Rich, M. S. Link, J. D. Schwartz, D. R. Gold, P. Koutrakis, R. L. Verrier, and M. A. Mittleman, "Particulate Air Pollution and Nonfatal Cardiac Events. Part II. Association of Air Pollution with Confirmed Arrhythmias Recorded by Implanted Defibrillators," *Research Reports, Health Effects Institute* 124 (2005): 83–126, discussion 127–148.

36. Mark S. Link, Heike Luttmann-Gibson, Joel Schwartz, Murray A. Mittleman, Benjamin Wessler, Diane R. Gold, Douglas W. Dockery, and Francine Laden, "Acute Exposure to Air Pollution Triggers Atrial Fibrillation," *Journal of the American College of Cardiology* 62, no. 9 (2013): 816–825.

37. C. Arden Pope, D. J. Eatough, D. R. Gold, Y. Pang, K. R. Nielsen, P. Nath, R. L. Verrier, and R. E. Kanner, "Acute Exposure to Environmental Tobacco Smoke and Heart Rate Variability," *Environmental Health Perspectives* 109, no. 7 (2001): 711–716.

38. Nicky Pieters, Michelle Plusquin, Bianca Cox, Michal Kicinski, Jaco Vangronsveld, and Tim S. Nawrot, "An Epidemiological Appraisal of the Association between Heart Rate Variability and Particulate Air Pollution: A Meta-Analysis" *Heart* 98, no. 15 (2012): 1127–1135.

39. Sarah Rajkumar, Arno Schmidt-Trucksäss, Gregory A. Wellenius, Georg F. Bauer, Cong Khanh Huynh, Alexander Moeller, and Martin Röösli, "The Effect of Workplace Smoking Bans on Heart Rate Variability and Pulse Wave Velocity of Non-Smoking Hospitality Workers," *International Journal of Public Health* 59, no. 4 (2014): 577–585.

40. Paul M. Ridker, "High-Sensitivity C-Reactive Protein," *Circulation* 103, no. 13 (2001): 1813–1818.

41. A. Nemmar, P. H. M. Hoet, B. Vanquickenborne, D. Dinsdale, M. Thomeer, M. F. Hoylaerts, H. Vanbilloen, L. Mortelmans, and B. Nemery, "Passage of Inhaled Particles into the Blood Circulation in Humans," *Circulation* 105, no. 4 (2002): 411–414.

42. G. Oberdörster, Z. Sharp, V. Atudorei, A. Elder, R. Gelein, W. Kreyling, and C. Cox, "Translocation of Inhaled Ultrafine Particles to the Brain," *Inhalation Toxicology* 16, nos. 6–7 (2004): 437–445.

43. Robert D. Brook, Sanjay Rajagopalan, C. Arden Pope, Jeffrey R. Brook, Aruni Bhatnagar, Ana V. Diez-Roux, Fernando Holguin, et al., "Particulate Matter Air Pollution and Cardiovascular Disease," *Circulation* 121, no. 21 (2010): 2331–2378; Sanjay Rajagopalan, Sadeer G. Al-Kindi, and Robert D. Brook, "Air Pollution and Cardiovascular Disease," *Journal of the American College of Cardiology* 72, no. 17 (2018): 2054–2070; Graham H. Bevan, Sadeer G. Al-Kindi, Robert D. Brook, Thomas Münzel, and Sanjay Rajagopalan, "Ambient Air Pollution and Atherosclerosis," *Arteriosclerosis, Thrombosis, and Vascular Biology* 41, no. 2a (2021): 628–637; Graham H. Bevan, Sadeer G. Al-Kindi, Robert Brook, and Sanjay Rajagopalan, "Ambient Air Pollution and Atherosclerosis: Recent Updates," *Current Atherosclerosis Reports* 23, no. 10 (2021), https://doi.org/10.1007/s11883-021-00958-9.

44. Graham H. Bevan, Sadeer G. Al-Kindi, Robert Brook, and Sanjay Rajagopalan, "Ambient Air Pollution and Atherosclerosis: Recent Updates," *Current Atherosclerosis Reports* 23, no. 10 (2021), https://doi.org/10.1007/s11883-021-00958-9.

45. C. Arden Pope, Richard T. Burnett, George D. Thurston, Michael J. Thun, Eugenia E. Calle, Daniel Krewski, and John J. Godleski, "Cardiovascular Mortality and Long-Term Exposure to Particulate Air Pollution," *Circulation* 109, no. 1 (2004): 71–77.

46. Stephan F. van Eeden, Adam Yeung, Kevin Quinlam, and James C. Hogg, "Systemic Response to Ambient Particulate Matter: Relevance to Chronic Obstructive Pulmonary Disease," *Proceedings of the American Thoracic Society* 2, no. 1 (2005): 61–67.

47. Don D. Sin and S. F. Paul Man, "Chronic Obstructive Pulmonary Disease as a Risk Factor for Cardiovascular Morbidity and Mortality," *Proceedings of the American Thoracic Society* 2, no. 1 (2005): 8–11; Don D. Sin, LieLing Wu, and S. F. Man, "The Relationship between Reduced Lung Function and Cardiovascular Mortality," *Chest* 127, no. 6 (2005): 1952–1959.

48. Tatsushi Suwa, James C. Hogg, Kevin B. Quinlan, Akira Ohgami, Renaud Vincent, and Stephan F. van Eeden, "Particulate Air Pollution Induces Progression of Atherosclerosis," *Journal of the American College of Cardiology* 39, no. 6 (2002): 935–942; van Eeden, Yeung, Quinlam, and Hogg, "Systemic Response to Ambient Particulate Matter."

49. Qinghua Sun, Aixia Wang, Ximei Jin, Alex Natanzon, Damon Duquaine, Robert D. Brook, Juan-Gilberto S. Aguinaldo, Zahi A. Fayad, Valentin Fuster, et. al, "Long-Term Air Pollution Exposure and Acceleration of Atherosclerosis and Vascular Inflammation in an Animal Model," *JAMA* 294, no. 23 (2005): 3003–3010.

50. Nino Künzli, Michael Jerrett, Wendy J. Mack, Bernardo Beckerman, Laurie LaBree, Frank Gilliland, Duncan Thomas, John Peters, and Howard N. Hodis, "Ambient Air Pollution and Atherosclerosis in Los Angeles," *Environmental Health Perspectives* 113, no. 2 (2005): 201–206.

51. Amy H. Auchincloss, Ana V. Diez Roux, J. Timothy Dvonch, Patrick L. Brown, R. Graham Barr, Martha L. Daviglus, David C. Goff, Joel D. Kaufman, and Marie S. O'Neill, "Associations between Recent Exposure to Ambient Fine Particulate Matter and Blood Pressure in the Multi-Ethnic Study of Atherosclerosis (MESA)," *Environmental Health Perspectives* 116, no. 4 (2008): 486–491.

52. Sara D. Adar, Lianne Sheppard, Sverre Vedal, Joseph F. Polak, Paul D. Sampson, Ana V. Diez Roux, Matthew Budoff, et al., "Fine Particulate Air Pollution and the Progression of Carotid Intima-Medial Thickness: A Prospective Cohort Study from the Multi-Ethnic Study of Atherosclerosis and Air Pollution," *PLoS Medicine* 10, no. 4 (2013): e1001430.

53. Joel D. Kaufman, Sara D. Adar, R. Graham Barr, Matthew Budoff, Gregory L. Burke, Cynthia L. Curl, Martha L. Daviglus, et al., "Association between Air Pollution and Coronary Artery Calcification within Six Metropolitan Areas in the USA (the Multi-Ethnic Study of Atherosclerosis and Air Pollution): A Longitudinal Cohort Study," *Lancet* 388, no. 10045 (2016): 696–704, 696.

54. Kaufman et al., "Association between Air Pollution and Coronary Artery Calcification," 696.

55. Mohammad Hashim Jilani, Bridget Simon-Friedt, Tamer Yahya, Ali Younas Khan, Syed Z. Hassan, Bita Kash, Ron Blankstein, et al., "Associations between Particulate Matter Air Pollution, Presence and Progression of Subclinical Coronary and Carotid Atherosclerosis: A Systematic Review," *Atherosclerosis* 306 (2020): 22–32.

56. Hongbing Xu, Tong Wang, Shengcong Liu, Robert D. Brook, Baihuan Feng, Qian Zhao, Xiaoming Song, et al., "Extreme Levels of Air Pollution Associated with Changes in Biomarkers of Atherosclerotic Plaque Vulnerability and Thrombogenicity in Healthy Adults," *Circulation Research* 124, no. 5 (2019): e30–e43; Seokhun Yang, Seung-Pyo Lee, Jun-Bean Park, Heesun Lee, Si-Hyuck Kang, Sang-Eun Lee, Juyong Brian Kim, Su-Yeon Choi, Yong-Jin Kim, and Hyuk-Jae Chang, "PM$_{2.5}$ Concentration in the Ambient Air Is a Risk Factor for the Development of High-Risk Coronary Plaques," *European Heart Journal—Cardiovascular Imaging* 20, no. 12 (2019): 1355–1364; Rocco A. Montone, Massimiliano Camilli, Michele Russo, Claudio Termite, Giulia La Vecchia, Giulia Iannaccone, Riccardo Rinaldi, et al., "Air Pollution and Coronary Plaque Vulnerability and Instability," *JACC: Cardiovascular Imaging* 15, no. 2 (2022): 325–342.

57. Robert D. Brook, Jeffrey R. Brook, Bruce Urch, Renaud Vincent, Sanjay Rajagopalan, and Frances Silverman, "Inhalation of Fine Particulate Air Pollution and Ozone Causes Acute Arterial Vasoconstriction in Healthy Adults," *Circulation* 105, no. 13 (2002): 1534–1536.

58. Thomas Münzel, Tommaso Gori, Sadeer Al-Kindi, John Deanfield, Jos Lelieveld, Andreas Daiber, and Sanjay Rajagopalan, "Effects of Gaseous and Solid Constituents of Air Pollution on Endothelial Function," *European Heart Journal* 39, no. 38 (2018): 3543–3550.

59. Bevan, Al-Kindi, Brook, and Rajagopalan, "Ambient Air Pollution and Atherosclerosis: Recent Updates."

60. C. Arden Pope, Aruni Bhatnagar, James P. McCracken, Wesley Abplanalp, Daniel J. Conklin, and Timothy O'Toole, "Exposure to Fine Particulate Air Pollution Is Associated with Endothelial Injury and Systemic Inflammation," *Circulation Research* 119, no. 11 (2016): 1204–1214.

61. Brook, Rajagopalan, Pope, et al., "Particulate Matter Air Pollution and Cardiovascular Disease."

62. David E. Newby, Pier M. Mannucci, Grethe S. Tell, Andrea A. Baccarelli, Robert D. Brook, Ken Donaldson, Francesco Forastiere, et al., "Expert Position Paper on Air Pollution and Cardiovascular Disease," *European Heart Journal* 36, no. 2 (2015): 83–93; Rajagopalan, Al-Kindi, and Brook, "Air Pollution and Cardiovascular Disease"; Münzel, Gori, Al-Kindi, et al., "Effects of Gaseous and Solid Constituents of Air Pollution"; Bevan, Al-Kindi, Brook, et al., "Ambient Air Pollution and Atherosclerosis."

CHAPTER 11

1. Swiss RE, "Our Approach," 2023, https://www.swissre.com/about-us/our-approach .html.

2. Douglas W. Dockery and C. Arden Pope, "Lost Life Expectancy Due to Air Pollution in China," *Risk Dialogue Magazine* 17 (2014): 2–48, https://www.swissre.com /dam/jcr:ca35b02a-dd17-49bf-99ef-076369cf7ff7/CGD_RDS_Health_Risk_Factors _Web.pdf; C. A. Pope and D. W. Dockery, "Air Pollution and Life Expectancy in

China and Beyond," *Proceedings of the National Academy of Sciences of the United States of America (PNAS)* 110, no. 32 (2013): 12861–12862.

3. GBD 2019 Risk Factors Collaborators, "Global Burden of 87 Risk Factors in 204 Countries and Territories, 1990–2019: A Systematic Analysis for the Global Burden of Disease Study 2019," *Lancet* 396, no. 10258 (2020): 1223–1249; WHO, "Household Air Pollution and Health: Fact Sheet," World Health Organization, September 22, 2021, https://www.who.int/news-room/fact-sheets/detail/household-air-pollution -and-health; Health Effects Institute, *State of Global Air 2020*, Special Report (Boston: Health Effects Institute, 2020), https://www.stateofglobalair.org; Aaron J. Cohen, Michael Brauer, Richard Burnett, H. Ross Anderson, Joseph Frostad, Kara Estep, Kalpana Balakrishnan, et al., "Estimates and 25-Year Trends of the Global Burden of Disease Attributable to Ambient Air Pollution: An Analysis of Data from the Global Burden of Diseases Study 2015," *Lancet* 389, no. 10082 (2017): 1907–1918.

4. J. Lelieveld, K. Klingmüller, A. Pozzer, R. T. Burnett, A. Haines, and V. Ramanathan, "Effects of Fossil Fuel and Total Anthropogenic Emission Removal on Public Health and Climate," *Proceedings of the National Academy of Sciences* 116, no. 15 (2019): 7192–7197; Richard Burnett, Hong Chen, Mieczysław Szyszkowicz, Neal Fann, Bryan Hubbell, C. Arden Pope, Joshua S. Apte, et al., "Global Estimates of Mortality Associated with Long-Term Exposure to Outdoor Fine Particulate Matter," *Proceedings of the National Academy of Sciences* 115, no. 38 (2018): 9592–9597; Karn Vohra, Alina Vodonos, Joel Schwartz, Eloise A. Marais, Melissa P. Sulprizio, and Loretta J. Mickley, "Global Mortality from Outdoor Fine Particle Pollution Generated by Fossil Fuel Combustion: Results from GEOS-Chem," *Environmental Research* 195 (2021): 110754.

5. Graham H. Bevan, Sadeer G. Al-Kindi, Robert D. Brook, Thomas Münzel, and Sanjay Rajagopalan, "Ambient Air Pollution and Atherosclerosis," *Arteriosclerosis, Thrombosis, and Vascular Biology* 41, no. 2a (2021): 628–637.

6. Gourab Choudhury and William MacNee, "Role of Inflammation and Oxidative Stress in the Pathology of Ageing in COPD: Potential Therapeutic Interventions," *COPD: Journal of Chronic Obstructive Pulmonary Disease* 14, no. 1 (2016): 122–135.

7. Michael Greenstone and Claire Quig Fan, "Introducing the Air Quality Life Index," Energy Policy Institute at the University of Chicago (EPIC), November 2018, https://aqli.epic.uchicago.edu/wp-content/uploads/2018/11/AQLI-Report.111918-2 .pdf.

8. Avraham Ebenstein, Maoyong Fan, Michael Greenstone, Guojun He, and Maigeng Zhou, "New Evidence on the Impact of Sustained Exposure to Air Pollution on Life Expectancy from China's Huai River Policy," *Proceedings of the National Academy of Sciences* 114, no. 29 (2017): 10384–10389.

9. Joshua S. Apte, Michael Brauer, Aaron J. Cohen, Majid Ezzati, and C. Arden Pope, "Ambient PM$_{2.5}$ Reduces Global and Regional Life Expectancy," *Environmental Science & Technology Letters* 5, no. 9 (2018): 546–551.

10. U.S. EPA, "The Benefits and Costs of the Clean Air Act, 1990 to 2020," U.S. Environmental Protection Agency Office of Air and Radiation. April 2011, https://www.epa.gov/sites/default/files/2015-07/documents/fullreport_rev_a.pdf.

11. B. Larsen, "Benefits and Costs of the Air Pollution: Targets for the Post-2015 Development Agenda," Air Pollution Assessment Paper, Post-2015 Consensus Project (Copenhagen: Copenhagen Consensus Center, 2014), https://www.copenhagenconsensus.com/sites/default/files/air_pollution_assessment_-_larsen.pdf.

12. World Bank and IHME (Institute for Health Metrics and Evaluation), "The Cost of Air Pollution: Strengthening the Economic Case for Action" (Washington, DC: World Bank, 2016).

13. OECD (Organisation for Economic Co-operation and Development), *The Economic Consequences of Outdoor Air Pollution* (Paris: OECD, 2016).

14. P. J. Landrigan, R. Fuller, N. J. R. Acosta, O. Adeyi, R. Arnold, N. N. Basu, A. B. Baldé, R. Bertollini, et al., "The Lancet Commission on Pollution and Health," *Lancet* 39, no. 10119 (2018): 462–512.

15. World Bank, "The Global Health Cost of Ambient $PM_{2.5}$ Air Pollution" (Washington, DC: World Bank, 2020), http://hdl.handle.net/10986/35721; World Bank, "The Global Health Cost of $PM_{2.5}$ Air Pollution: A Case for Action beyond 2021," International Development in Focus (Washington, DC: World Bank, 2022), doi:10.1596/978-1-4648-1816-5.

16. World Bank, "The Global Health Cost of $PM_{2.5}$ Air Pollution: A Case for Action beyond 2021."

17. World Bank, "The Global Health Cost of $PM_{2.5}$ Air Pollution: A Case for Action beyond 2021."

18. GBD 2019 Risk Factors Collaborators, "Global Burden."

19. World Bank, "The Global Health Cost of $PM_{2.5}$ Air Pollution: A Case for Action beyond 2021."

20. U.S. EPA, "The Benefits and Costs of the Clean Air Act, 1970 to 1990," prepared for U.S. Congress by U.S. Environmental Protection Agency, October 1997, https://www.epa.gov/sites/default/files/2015-06/documents/contsetc.pdf; U.S. EPA, "The Benefits and Costs of the Clean Air Act, 1990 to 2010," EPA Report to Congress, November 1999, https://www.epa.gov/sites/default/files/2015-07/documents/fullrept.pdf.

21. Office of Management and Budget, *2007 Report to Congress on the Benefits and Costs of Federal Regulations and Unfunded Mandates on State, Local, and Tribal Entities*, Office of Management and Budget Office of Information and Regulatory Affairs, Report to Congress, https://www.whitehouse.gov/wp-content/uploads/legacy_drupal_files/omb/assets/OMB/inforeg/2007_cb/2007_cb_final_report.pdf.

22. U.S. EPA, "Our Nation's Air: Trends through 2022," U.S. Environmental Protection Agency, 2023, https://gispub.epa.gov/air/trendsreport/2023/#air_trends.

23. U.S. EPA, "The Benefits and Costs of the Clean Air Act, 1970 to 1990"; U.S. EPA, Integrated Science Assessment (ISA) for Particulate Matter (Final Report, Dec. 2019).

24. Harald Fuller-Bennett and Iris Velez, "Woodsy Owl at 40," *Forest History Today*, spring 2012, https://foresthistory.org/wp-content/uploads/2016/12/2012-Spring _Woodsy-Owl-at-40.pdf.

25. David Michaels, *Doubt Is Their Product: How Industry's Assault on Science Threatens Your Health* (Oxford: Oxford University Press, 2008); David Michaels, *The Triumph of Doubt: Dark Money and the Science of Deception* (Oxford: Oxford University Press, 2020); Naomi Oreskes and Erik M. Conway, *Merchants of Doubt: How a Handful of Scientists Obscured the Truth on Issues from Tobacco Smoke to Global Warming* (New York: Bloomsbury Press, 2010).

26. Richard Conniff, "The Political History of Cap and Trade," *Smithsonian Magazine*, August 2009, https://www.smithsonianmag.com/science-nature/the-political -history-of-cap-and-trade-34711212.

27. Climate Leadership Council, *The Four Pillars of the Carbon Dividends Plan*, https:// clcouncil.org/our-solution.

CHAPTER 12

1. Rebecca Fairley Raney, "Robert Phalen Tests Our Modern Air," Member Spotlight, American Association for the Advancement of Science, July 30, 2012, 1; for a presentation of his views, see Robert F. Phalen, "A Toxicologist's Views on PM and Human Health," presented at the 35th Annual Meeting of DDP (Doctors for Disaster Preparedness), August 12, 2017, https://www.youtube.com/watch?v=uSMRA-6PIL8.

2. Summer Meza, "Incoming EPA Adviser Thinks Air Is Too Clean," *Newsweek*, November 2, 2017.

3. Emma Howard, "'Modern Air Is a Little Too Clean': The Rise of Air Pollution Denial," *Unearthed*, November 11, 2017.

4. Robert F. Phalen, *The Particulate Air Pollution Controversy: A Case Study and Lessons Learned* (Boston: Kluwer Academic, 2002).

5. Raney, "Robert Phalen Tests Our Modern Air," 1.

6. WHO, "WHO Global Air Quality Guidelines: Particulate Matter (PM2.5 And PM10), Ozone, Nitrogen Dioxide, Sulfur Dioxide and Carbon Monoxide," September 22, 2021.

7. American Lung Association, *State of the Air*, 2022, https://www.lung.org/research /sota.

8. David E. Newby, Pier M. Mannucci, Grethe S. Tell, Andrea A. Baccarelli, Robert D. Brook, Ken Donaldson, Francesco Forastiere, et al., "Expert Position Paper on Air Pollution and Cardiovascular Disease," *European Heart Journal* 36, no. 2 (2015): 83–93.

9. Sanjay Rajagopalan, Michael Brauer, Aruni Bhatnagar, Deepak L. Bhatt, Jeffrey R. Brook, Wei Huang, Thomas Münzel, David Newby, Jeffrey Siegel, and Robert D. Brook, "Personal-Level Protective Actions against Particulate Matter Air Pollution

Exposure: A Scientific Statement from the American Heart Association," *Circulation* 142, no. 23 (2020), https://doi.org/10.1161/cir.0000000000000931.

10. Sanjay Rajagopalan, Sadeer G. Al-Kindi, and Robert D. Brook, "Air Pollution and Cardiovascular Disease," *Journal of the American College of Cardiology* 72, no. 17 (2018): 2054–2070.

11. IARC, "Air Pollution and Cancer," IARC Scientific Publication No. 161 (Lyon: International Agency for Research on Cancer, 2013), https://www.iarc.who.int/news-events/iarc-scientific-publication-no-161-air-pollution-and-cancer-as-an-e-book.

12. Michelle C. Turner, Zorana J. Andersen, Andrea Baccarelli, W. Ryan Diver, Susan M. Gapstur, C. Arden Pope, Diddier Prada, Jonathan Samet, George Thurston, and Aaron Cohen, "Outdoor Air Pollution and Cancer: An Overview of the Current Evidence and Public Health Recommendations," *CA: A Cancer Journal for Clinicians* 70, no. 6 (2020): 460–479.

13. Academy of Science of South Africa, Brazilian Academy of Sciences, German National Academy of Sciences Leopoldina, U.S. National Academy of Medicine, and U.S. National Academy of Sciences, "Air Pollution and Health—A Science-Policy Initiative," *Annals of Global Health* 85, no. 1 (2019): 140, 1.

14. U.S. EPA, Integrated Science Assessment (ISA) for Particulate Matter (Final Report, Dec. 2019).

15. Health Effects Institute, *State of Global Air 2020*, Special Report (Boston: Health Effects Institute, 2020), https://www.stateofglobalair.org.

16. WHO, "Ambient (Outdoor) Air Pollution: Fact Sheet, World Health Organization," September 22, 2021, https://www.who.int/news-room/fact-sheets/detail/ambient-(outdoor)-air-quality-and-health.

17. Jonathan Gruber, *Public Finance and Public Policy* (New York: Worth Publishers, 2019).

18. Eduardo Medina and Michael Levenson, "Amtrak Train Hits Truck in Missouri, Killing 4 and Injuring Dozens," *New York Times*, June 27, 2022, https://www.nytimes.com/2022/06/27/us/amtrak-truck-crash-missouri.html.

19. David Michaels, *Doubt Is Their Product: How Industry's Assault on Science Threatens Your Health* (Oxford: Oxford University Press, 2008); David Michaels, *The Triumph of Doubt: Dark Money and the Science of Deception* (Oxford: Oxford University Press, 2020); Naomi Oreskes and Erik M. Conway, *Merchants of Doubt: How a Handful of Scientists Obscured the Truth on Issues from Tobacco Smoke to Global Warming* (New York: Bloomsbury Press, 2010).

20. Steven J. Milloy, *Scare Pollution: Why and How to Fix the EPA* (United States: Bench Press, 2016).

21. Robert N. Proctor, *Golden Holocaust: Origins of the Cigarette Catastrophe and the Case for Abolition* (Berkeley: University of California Press, 2012).

22. C. Arden Pope, Maureen Cropper, Jay Coggins, and Aaron Cohen, "Health Benefits of Air Pollution Abatement Policy: Role of the Shape of the Concentration–Response

Function," *Journal of the Air & Waste Management Association* 65, no. 5 (2014): 516–522.

23. Kirk R. Smith, Michael Jerrett, H. Ross Anderson, Richard T. Burnett, Vicki Stone, Richard Derwent, Richard W. Atkinson, et al., "Public Health Benefits of Strategies to Reduce Greenhouse-Gas Emissions: Health Implications of Short-Lived Greenhouse Pollutants," *Lancet* 374, no. 9707 (2009): 2091–2103; Rajagopalan, Al-Kindi, and Brook, "Air Pollution and Cardiovascular Disease"; Frederica Perera and Kari Nadeau, "Climate Change, Fossil-Fuel Pollution, and Children's Health," *New England Journal of Medicine* 386, no. 24 (2022): 2303–2314.

24. U.S. EPA, "Fact Sheet: Overview of the Clean Power Plan. U.S. Environmental Protection Agency," https://19january2017snapshot.epa.gov/cleanpowerplan/fact-sheet-overview-clean-power-plan_.html.

25. C. Boyden Gray, "EPA's Use of Co-Benefits," *Engage* 16, no. 2 (2015): 31–33, https://fedsoc.org/commentary/publications/epa-s-use-of-co-benefits.

26. Charles T. Driscoll, Jonathan J. Buonocore, Jonathan I. Levy, Kathleen F. Lambert, Dallas Burtraw, Stephen B. Reid, Habibollah Fakhraei, and Joel Schwartz, "U.S. Power Plant Carbon Standards and Clean Air and Health Co-Benefits," *Nature Climate Change* 5, no. 6 (2015): 535–540.

27. James Gustave Speth, *They Knew: The U.S. Federal Government's Fifty-Year Role in Causing the Climate Crisis* (Cambridge, MA: MIT Press, 2022).

28. C. Arden Pope, "Air Pollution and Health—Good News and Bad," *New England Journal of Medicine* 351, no. 11 (2004): 1132–1134.

29. Austin Bradford Hill, "The Environment and Disease: Association or Causation?," *Proceedings of the Royal Society of Medicine.* 58, no. 5 (1965): 295–300.

INDEX